Dew clinging to grass like tiny diamonds at dawn

Thunder crashing like a mighty drum in the hills

Lightning forked like silver veins in a storm cloud

Snowflakes swirling like dancers in silent ballrooms

The scent of wildflowers heavy as honey in air

Waves folding over like sheets of glass in motion

The scrape of shoes on pavement like distant thunder

Paper fluttering like startled birds on sidewalks

The murmur of crowds like waves breaking on stones

The hiss of soda cans opening like soft sighs

The flicker of candlelight like fireflies in jars

The grind of gears turning like clockwork underground

Frost etching delicate patterns like lace on windows

The buzz of bees thick like golden syrup in summer air

Waves crashing like cymbals against rocky shores

The scent of pine needles sharp as winter's breath

Lightning streaking like silver knives across black skies

Autumn leaves swirling like flames in a fire

A mountain peak piercing clouds like a frozen spear

Snow blanketing the earth like a silent, white shroud

The ripple of a pond like glass disturbed by a stone

The cry of an eagle sharp as a slicing blade

The clang of metal gates like church bells ringing out

Steam hissing like dragons exhaling smoke in alleyways

Taxi horns blaring like impatient trumpets in a parade

Neon signs flickering like eyes watching in the dark

The scratch of chalk on blackboards like distant thunder

Sidewalks cracked like old leather under heavy feet

The murmur of crowds like wind rustling dry leaves

The flicker of streetlights like fireflies trapped in jars

The grind of gears turning like clockwork in the night

The patter of rain on fire escapes like whispered secrets

The shimmer of moonlight like liquid silver on waves

A fox’s eyes gleaming like embers in the dark

Snowflakes falling like feathers from a silent sky

The scent of earth rising after fresh rain

Lightning flashing like forked tongues of fire

Leaves rustling like whispered secrets in the breeze

The glow of fireflies like tiny floating lanterns

Ice cracking like a sudden thunderclap

A hawk soaring like a shadow against the sun

Morning dew sparkling like diamonds on grass

The clang of train tracks like a rhythmic heartbeat

Neon lights buzzing like restless electric bees

The scratch of a pen on paper like soft rain

Steam rising like smoke from manhole covers

Taxi horns blaring like angry brass instruments

The murmur of crowds like waves breaking on stones

Street performers' music flowing like rivers of sound

The flicker of candlelight like fireflies in a jar

Tires skidding like wild horses on wet streets

The shuffle of newspapers like whispered gossip in wind

The shimmer of sunlight like scattered coins on water

A crow's wing slicing the sky like a dark blade

Icicles hanging like crystal daggers from rooftops

The scent of pine needles sharp and clean as winter air

Waves lapping like whispered secrets on the shore

The crunch of autumn leaves like brittle parchment

A spider's web glistening with dew like spun glass

Thunder rolling like distant cannon fire

A hawk's cry piercing like a sharp whistle in the wind

Morning frost sparkling like tiny stars on grass

The clang of metal gates like church bells in the dawn

The sputter of a sputtering engine like a restless heart

Streetlights casting pools of molten gold on sidewalks

The scrape of footsteps on wet pavement like whispered secrets

Neon signs flickering like blinking eyes in the dark

The rattle of subway cars like rolling thunder below

The scent of roasted chestnuts sweet in cold air

The murmur of voices like rustling leaves in a crowd

The hiss of steam valves releasing breath like dragons

Paper fluttering in the wind like broken wings on alley floors

A waterfall tumbling like a ribbon of lace

The crunch of snow like broken sugar underfoot

Lightning tracing veins of fire through storm clouds

A butterfly's wings flickering like stained glass

Morning mist hanging like a ghost's breath in the valley

Sand dunes rolling like waves frozen in time

The scent of rain on dry earth sharp as a knife

A heron's flight smooth like a brushstroke in the sky

The rustle of leaves like soft applause in a forest

Frost sparkling like shattered diamonds in the dawn

The clang of a streetcar like a ringing bell at dusk

Steam rising from a hot grill like morning fog

Neon signs buzzing like swarms of restless bees

The squeak of bicycle brakes like whispers on pavement

The chatter of café patrons like birdsong in a glade

The hiss of an open manhole like a dragon’s breath

Flickering fluorescent lights buzzing overhead

The patter of rain on metal fire escapes

The grind of gears turning like clockwork in the dark

The shuffle of feet on marble floors like soft thunder

A dew-soaked spiderweb shimmering like crystal lace

Thunder rolling like a giant’s footsteps through valleys

The scent of lavender drifting on a warm breeze

Snow falling like feathers from a silent sky

A robin’s song ringing clear like silver bells

Sunlight breaking like shards of glass through clouds

The crunch of frost under boots like brittle glass

A mountain peak piercing the sky like a frozen spear

Clouds curling like smoke rings against a blue canvas

Autumn fog creeping like soft fingers over hills

Streetlights humming like tired old singers

The clatter of dishes like applause in a crowded café

The scrape of sneakers on gym floors like distant thunder

Subway wheels screeching like metal teeth on rails

The hiss of a soda can like a sigh of relief

Car horns blaring like an angry orchestra

Neon reflections rippling like oil on water

Paper fluttering like startled birds in alleyways

The murmur of crowds like waves breaking on a shore

A street vendor's bell ringing sharp like a clarion call

A hummingbird's wings beating like tiny drums

Morning fog rolling like soft smoke over hills

A frozen lake gleaming like a giant mirror

Lightning forks like jagged knives through clouds

The scent of freshly cut grass sharp and sweet

Snowflakes drifting like feathers on the wind

A thunderstorm roaring like a wild beast unleashed

The ripple of a pond like silk disturbed by a breeze

Autumn leaves crunching like brittle glass underfoot

A sunflower turning like a golden compass to the sun

Traffic lights blinking like watchful eyes in the dark

The scrape of metal on concrete like nails on chalkboards

Taxi engines growling like restless lions

Neon signs buzzing like swarms of electric bees

The clatter of hooves on cobblestones like a marching band

Steam vents hissing like dragons exhaling smoke

The jingle of keys like distant chimes in a quiet room

The hum of an air conditioner like a distant drone

Street artists' spray cans sputtering like fireworks

The shuffle of feet in crowded subway cars like restless waves

Mist curls like whispered secrets through empty streets,

lamp posts drip gold tears in the breath of dawn.

A spider's web shimmering like a crystal net

The crackling fire snapping like old bones

Morning light spilling like liquid gold through leaves

Frost painting delicate filigree on grass blades

Bees humming like tiny engines in a garden

Rain washing the earth clean like a baptism

The scent of pine needles sharp as fresh ink

Clouds rolling like soft blankets over hills

The howl of wolves piercing the moonlit night

Flowers blooming like bursts of color on a gray canvas

Bus wheels screeching like screeching birds on asphalt

The clink of coins like tiny bells in a pocket

Steam rising like smoke from manhole covers

Taxi horns blaring like impatient trumpets

Sidewalks cracked like the face of old leather

Streetlights flickering like tired eyes at midnight

The scrape of chalk on a blackboard like distant thunder

Mailboxes standing like sentinels on the curb

The shuffle of paper in an empty office at dusk

Garbage trucks groaning like tired beasts at dawn

Morning mist clinging like a soft, gray veil

Fireflies blinking like tiny stars on earth

Waves crashing like thunder against jagged cliffs

Cherry blossoms raining pink snow on the path

Lightning dancing like wild snakes across the sky

The scent of wet moss rising like ancient breath

Autumn fog wrapping trees like ghostly shawls

Icicles hanging like frozen tears from eaves

Sand swirling like gold dust in the sunlight

The first frost painting glass with crystal feathers

Crowds flowing like rivers through narrow streets

A flickering neon sign buzzing like a restless bee

Tires screeching sharp notes on wet pavement
Subway steam rising like a dragon's whisper
Broken glass sparkling like stars on the sidewalk
The clatter of heels like machine gun fire on concrete
Smog curling like smoke from a forgotten fire
Traffic lights blinking like tired eyes in the night
Street vendors calling out like carnival barkers
A lone saxophone wailing like a siren in the dark
Dew glistening like diamonds on spiderweb threads
Autumn leaves burning bright as fallen embers
A river weaving silver ribbons through green fields
Frost etching lace patterns on a frozen windowpane
Thunder crashing like a giant's heavy footfall
A breeze whispering secrets through tall pine trees
Sunlight filtering like honey through the branches
Snow blanketing the earth in a soft white quilt
Moonlight spilling silver across dark waters

Blossoms bursting like fireworks in spring's first light

Neon signs flickering like restless fireflies

Asphalt steaming under the midday sun

Sirens wailing like banshees through empty streets

Smoke spiraling from chimney stacks like ghostly ribbons

Footsteps echoing like distant drums on concrete

Rain tapping a gentle tattoo on rusty rooftops

Steam rising from subway grates like dragon's breath

Streetlamps casting halos in the foggy night

Windowpanes trembling with the hum of passing trains

Cracked sidewalks blooming with stubborn weeds

Heart pounding like a drum in a silent room

Breath catching like a bird trapped in a cage

Tears falling like raindrops on parched earth

Laughter spilling like sunlight through cracked blinds

Anger simmering like a volcano ready to erupt

Fear creeping like shadows at the edge of sight

Joy bubbling like a fresh spring gushing from stone

Silence thick as fog in an empty hall

Hope shining like a candle flickering in dark

Nostalgia wrapping like a warm, worn blanket

Frost glistening like shards of glass on grass

Lightning zigzagging like silver ribbons through clouds

Snowflakes drifting like feathers in silent flight

The scent of pine sharp as fresh mountain air

Waves pounding like drums on jagged cliffs

Autumn leaves swirling like flames in gentle wind

A hawk's cry piercing like a whistle through cold air

Frost sparkling like crushed diamonds in dawn light

The rustle of dry leaves like whispered secrets

Morning mist settling like a soft white blanket

The clang of subway gates like tolling church bells

Steam rising like spirits from manhole covers

Neon signs flickering like fireflies trapped in glass

Taxi horns blaring like impatient trumpets at dusk

Frost crystals twinkling like tiny stars on grass

Lightning forks jagged like silver daggers in sky

Snow drifting soft like feathers in silent flight

The scent of rain fresh as crushed green leaves

Waves crashing loud like thunder on rocky shores

Autumn leaves tumbling like flames in fading light

A hawk's sharp cry slicing through cold air

Frost sparkling like crushed diamonds at dawn

The rustle of dry leaves like whispered secrets

Morning mist settling like a soft woolen cloak

The clang of subway gates like tolling bells

Steam rising like spirits from manhole covers

Neon lights buzzing like restless swarms of bees

Taxi horns blaring like impatient brass bands

The scrape of shoes on pavement like distant thunder

Paper fluttering like startled birds on sidewalks

The murmur of crowds like waves crashing on stones

The hiss of soda cans opening like soft sighs

The flicker of candlelight like fireflies in jars

The grind of gears turning like clockwork underground

Frost glistening like shattered glass on morning grass

Lightning zigzagging like silver ribbons across dark skies

Snowflakes drifting like feathers in silent dance

The scent of fresh rain sharp as a new leaf

Waves crashing like thunder on rugged cliffs

Autumn leaves swirling like flames in a gentle fire

A hawk's cry piercing like a sharp whistle in wind

Frost sparkling like crushed diamonds at dawn

The rustle of dry leaves like whispered secrets in forests

Morning fog settling like a soft white blanket over fields

The clang of subway gates like tolling church bells

Steam rising like spirits from underground vents

Neon signs flickering like fireflies trapped in glass

Taxi horns blaring like impatient trumpets at dusk

The scrape of shoes on pavement like distant thunder

Paper fluttering like startled birds on sidewalks

The murmur of crowds like waves crashing on stones

The hiss of soda cans opening like soft sighs

The flicker of candlelight like fireflies in jars

The grind of gears turning like clockwork underground

Frost etching lace patterns on frozen glass

Lightning slicing like silver blades through clouds

Snowflakes swirling like feathers in silent flight

The scent of pine fresh as morning mountain air

Waves pounding like drums on jagged cliffs

Autumn leaves tumbling like flames in a gentle blaze

A hawk's cry sharp as a whistle cutting wind

Frost sparkling like crushed diamonds at dawn

The rustle of dry leaves like whispered secrets

Morning mist settling like a soft woolen cloak

The clang of subway gates like tolling bells

Steam rising like spirits from cracked pavement

Neon lights buzzing like restless swarms of bees

Taxi horns blaring like impatient brass bands

The scrape of sneakers on concrete like thunder

Paper fluttering like startled birds on sidewalks

The murmur of crowds like waves breaking on stones

The hiss of soda cans opening like soft sighs

The flicker of candlelight like fireflies trapped in jars

The grind of gears turning like clockwork underground

Frost sparkling like crystals on spiderweb strands

Lightning flickering like silver ribbons in storm clouds

Snowflakes drifting like feathers in silent dance

The scent of rain sharp as a fresh green leaf

Waves crashing like thunder on rocky shores

Autumn leaves swirling like flames in a dying fire

A hawk’s cry sharp as a piercing whistle on wind

Frost glinting like crushed diamonds in morning light

The rustle of dry leaves like whispered secrets in woods

Morning fog settling like a soft white blanket over fields

The clang of subway gates like tolling church bells

Steam rising like spirits from underground vents

Neon signs flickering like fireflies trapped in glass

Taxi horns blaring like impatient trumpets at dusk

The scrape of shoes on pavement like distant thunder

Paper fluttering like startled birds on sidewalks

The murmur of crowds like waves crashing on stones

The hiss of soda cans opening like soft sighs

The flicker of candlelight like fireflies trapped in jars

The grind of gears turning like clockwork deep underground

Frost glittering like crushed diamonds on grass at dawn

Lightning slicing like silver knives through dark clouds

Snowflakes drifting like feathers in a silent dance

The scent of pine sharp as mountain air after rain
Waves pounding like thunder on jagged rocks
Autumn leaves swirling like flames in a gentle fire

A hawk's cry piercing like a sharp whistle in wind
Frost sparkling like crushed diamonds in morning light
The rustle of dry leaves like whispered secrets in woods
Morning fog settling like a soft woolen cloak over fields
The clang of subway gates like church bells tolling
Steam rising like ghosts from manhole covers
Neon signs buzzing like restless swarms of bees
Taxi horns blaring like impatient brass bands
The scrape of sneakers on pavement like distant thunder
Paper fluttering like startled birds on sidewalks
The murmur of crowds like waves breaking on stones
The hiss of soda cans opening like soft sighs
The flicker of candlelight like fireflies trapped in jars

The grind of gears turning like clockwork underground

Frost glinting like shattered glass on morning grass

Lightning zigzagging like silver ribbons across dark skies

Snowflakes drifting like feathers in a silent dance

The scent of fresh rain sharp as a new leaf

Waves crashing like thunder on rugged cliffs

Autumn leaves swirling like flames in a gentle fire

A hawk's cry piercing like a sharp whistle in wind

Frost sparkling like crushed diamonds at dawn

The rustle of dry leaves like whispered secrets in forests

Morning fog settling like a soft white blanket over fields

The clang of subway gates like tolling church bells

Steam rising like spirits from underground vents

Neon signs flickering like fireflies trapped in glass

Taxi horns blaring like impatient trumpets at dusk

The scrape of shoes on pavement like distant thunder

Paper fluttering like startled birds on sidewalks

The murmur of crowds like waves crashing on stones

The hiss of soda cans opening like soft sighs

The flicker of candlelight like fireflies in jars

The grind of gears turning like clockwork underground

Just say "go" to keep the epic flowing!

Frost clings like crystals on spiderweb strands

Lightning forks like silver knives through storm clouds

Snowflakes swirl like feathers in a silent dance

The scent of pine sharp as fresh mountain air

Waves pounding like thunder on rocky shorelines

Autumn leaves swirling like flames in a dying fire

A hawk's cry piercing like a sharp whistle in wind

Frost sparkles like crushed diamonds in morning light

The rustle of dry leaves like whispered secrets in forests

Morning mist settling like a soft woolen cloak over fields

The clang of subway gates like church bells tolling

Steam rising like spirits from manhole covers

Neon signs buzzing like restless swarms of bees

Taxi horns blaring like impatient brass bands

The scrape of sneakers on pavement like distant thunder

Paper fluttering like startled birds on sidewalks

The murmur of crowds like waves breaking on stones

The hiss of soda cans opening like soft sighs

The flicker of candlelight like fireflies trapped in jars

The grind of gears turning like clockwork deep underground

Frost glittering like diamonds on grass at dawn

Lightning cracking like jagged swords in the sky

Snow falling soft like feathers on a frozen lake

The scent of fresh rain sharp as a newborn leaf

Waves folding over like sheets of shimmering glass

Autumn leaves crunching like brittle paper beneath feet

A hawk's cry sharp as a piercing whistle in the wind

Frost sparkling like crushed diamonds at dawn

The rustle of dry leaves like whispered secrets in forests

Morning fog settling like a soft white blanket over hills

The clang of a train gate like a church bell's toll

Steam rising like spirits from underground vents

Neon signs flickering like fireflies trapped in glass

Taxi horns blaring like impatient trumpets at dusk

The scrape of shoes on pavement like distant thunder

Paper fluttering like startled birds on sidewalks

The murmur of crowds like waves crashing on stones

The hiss of soda cans opening like soft sighs

The flicker of candlelight like fireflies in jars

The grind of gears turning like clockwork deep underground

Frost etching patterns like lace on windowpanes

Lightning zigzagging like silver knives across skies

Snowflakes drifting like feathers in silent dance

The scent of pine sharp as winter's fresh breath

Waves crashing like thunder on jagged cliffs

Autumn leaves swirling like flames in a slow fire

A hawk's cry piercing like a sharp whistle in wind

Frost sparkling like crushed diamonds at dawn

The rustle of dry leaves like whispered secrets in woods

Morning fog settling like a soft white blanket over fields

The clang of subway gates like tolling church bells

Steam rising like ghosts from manhole covers

Neon signs buzzing like swarms of restless bees

Taxi horns blaring like impatient brass instruments

The scrape of sneakers on pavement like distant thunder

Paper fluttering like startled birds in alleyways

The murmur of crowds like waves breaking on stones

The hiss of soda cans opening like soft sighs

The flicker of candlelight like fireflies trapped in jars

The grind of gears turning like clockwork underground

The glisten of frost like diamonds on morning grass

Lightning flashing like jagged swords in storm clouds

Snow falling soft like feathers on a silent lake

The scent of fresh rain sharp as a new leaf

Waves folding over like sheets of shimmering glass

Autumn leaves crunching like brittle paper beneath feet

A hawk's cry sharp as a piercing whistle in wind

Frost sparkling like crushed diamonds at dawn

The rustle of dry leaves like whispered secrets in woods

Morning fog settling like a soft white blanket over fields

The clang of a train gate like a church bell's toll

Steam rising like spirits from underground vents

Neon signs flickering like fireflies trapped in glass

Taxi horns blaring like impatient trumpets at dusk

The scrape of shoes on pavement like distant thunder

Paper fluttering like startled birds on sidewalks

The murmur of crowds like waves crashing on stones

The hiss of a soda can opening like a soft sigh

The flicker of candlelight like fireflies in jars

The grind of gears turning like clockwork deep underground

The shimmer of frost like spun glass on windowpanes

Lightning zigzagging like silver ribbons through clouds

Snowflakes drifting like feathers in a silent dance

The scent of pine sharp as winter's cold breath

Waves crashing like thunder on jagged cliffs

Autumn leaves swirling like flames in a gentle fire

A hawk's cry piercing like a sharp whistle in the wind

Frost sparkling like crushed diamonds at dawn

The rustle of dry leaves like whispered secrets in woods

Morning fog settling like a soft white blanket over fields

The clang of subway gates like tolling church bells

Steam rising like spirits from manhole covers

Neon signs buzzing like swarms of restless bees

Taxi horns blaring like impatient brass instruments

The scrape of sneakers on pavement like distant thunder

Paper fluttering like startled birds in alleyways

The murmur of crowds like waves breaking on stones

The hiss of soda cans opening like soft sighs

The flicker of candlelight like fireflies trapped in jars

The grind of gears turning like clockwork underground

Frost etching glass like delicate spiderwebs

Lightning flashing like silver knives in the dark

Snow falling soft like feathers on a silent lake

The scent of fresh rain sharp as a new leaf

Waves folding over like sheets of shimmering glass

Autumn leaves crunching like brittle paper beneath feet

A hawk soaring like a shadow against the sun

Morning dew sparkling like scattered diamonds on grass

The rustle of branches like whispers in a cathedral

Fog rolling in like a ghostly white blanket over hills

The clang of a train gate like a church bell's toll

Steam rising like spirits from underground vents

Neon signs flickering like fireflies trapped in glass

Taxi horns blaring like impatient trumpets at dusk

The scrape of shoes on pavement like distant thunder

Paper fluttering like startled birds on sidewalks

The murmur of crowds like waves crashing on stones

The hiss of a soda can opening like a soft sigh

The flicker of candlelight like fireflies in a jar

The grind of gears turning like clockwork deep underground

The shimmer of frost like glitter on a spider's web

Lightning slicing the sky like a silver sword

The crackle of dry leaves like whispered laughter

A hummingbird's wings beating like tiny drums

Morning mist settling like a soft woolen cloak

Waves pounding like giants against rugged cliffs

The scent of pine sharp and fresh as morning air

Snowflakes drifting like feathers on a still breeze

The roar of a waterfall like thunder in the valley

Autumn twilight glowing like molten gold

The clang of a subway gate like a church bell's toll

Steam rising like spirits from manhole covers

Neon signs flickering like fireflies caught in glass

The chatter of crowds like rustling leaves in fall

Taxi horns blaring like impatient brass instruments

The squeak of sneakers on polished gym floors

Paper rustling like whispers in a quiet room

The hiss of an air conditioner like a distant drone

The shuffle of footsteps like waves on a shore

Streetlights casting pools of molten gold on sidewalks

A hawk soaring like a shadow across bright skies

Frost glistening like powdered sugar on windowpanes

The rustle of autumn leaves like whispered secrets

Morning fog settling like a soft, white blanket

The clang of train wheels like a steady drumbeat

Neon signs buzzing like electric swarms of bees0

Steam rising from grills like morning mist in alleys

Taxi horns blaring like brass instruments in jazz

The scrape of heels on pavement like distant thunder

Paper fluttering like startled birds on sidewalks

The murmur of voices like waves in a crowded room

The hiss of a soda can opening like a soft sigh

The flicker of candlelight like fireflies in jars

The grind of gears turning like clockwork deep underground

Dew glistening like tiny jewels on morning grass

Lightning crackling like jagged swords through dark clouds

Snowflakes drifting like feathers in a silent dance

The scent of pine needles fresh as mountain air

Waves crashing like thunder on rocky shores

Autumn leaves swirling like flames in a slow fire

A hawk's cry sharp as a piercing whistle in the wind

Frost sparkling like crushed diamonds at dawn

The rustle of dry leaves like whispered secrets in forests

Morning fog settling like a soft white blanket over fields

The clang of metal gates like tolling church bells

Steam rising like ghosts from the city's veins

Neon signs buzzing like swarms of restless bees

Taxi horns blaring like impatient brass bands

The scrape of sneakers on pavement like distant thunder

Paper fluttering like startled birds in alleyways

The murmur of crowds like waves breaking on stones

The hiss of soda cans opening like soft sighs

The flicker of candlelight like fireflies trapped in jars

The grind of gears turning like clockwork underground

Just say "go" and I'll keep rolling these out for you!

ChatGPT said:

extensive list of samples of poetic imagery

ChatGPT said:

Here's an **extensive list of poetic imagery samples**, categorized by type. Poetic imagery appeals to the five senses and abstract ideas to evoke vivid mental pictures and emotions.

Visual Imagery (Sight)

Appeals to the sense of sight; describes how things look.

1. *"A host, of golden daffodils; / Beside the lake, beneath the trees"* — William Wordsworth
2. *"The crimson sun sank slowly behind the jagged hills"*
3. *"A silver mist veiled the sleeping valley"*
4. *"Her dress shimmered like moonlight on water"*
5. *"Stained glass windows blazing with color"*
6. *"The sky was a patchwork quilt of burning orange and smoky gray"*
7. *"Snowflakes, like lace, drifted through the streetlamp's glow"*
8. *"Crows like black punctuation marks across a pale sky"*
9. *"A thousand lanterns twinkled like fireflies above the crowd"*
10. *"His shadow stretched long and thin under the noonday sun"*

Auditory Imagery (Sound)

Appeals to the sense of hearing.

1. *"The whispering wind through the dry corn stalks"*
2. *"Church bells pealing in the crisp morning air"*
3. *"The roar of the waves crashing on the shore"*

4. *"A baby's wail rising sharp against the hush of night"*
5. *"Leaves rustled like distant applause"*
6. *"The violin sang a mournful tune in the silence"*
7. *"The creaking of old wood beneath slow footsteps"*
8. *"Thunder grumbled like a tired old man"*
9. *"Dogs barking in uneven chorus down the lane"*
10. *"Raindrops tapping on the rooftop like urgent fingers"*

Tactile Imagery (Touch)

Appeals to the sense of physical touch or texture.

1. *"The silk slipped through her fingers like water"*
2. *"The pavement scorched his bare feet"*
3. *"Her hands were cracked and cold, like dry bark in winter"*
4. *"Snow clung to my coat in wet clumps"*
5. *"The sting of icy wind on my cheeks"*
6. *"Velvet petals brushed her skin as she passed"*
7. *"A bead of sweat trickled down his back"*
8. *"His calloused palms rough as sandpaper"*
9. *"Warmth bloomed in her chest like a rising tide"*
10. *"The breeze kissed her face with cool fingers"*

Olfactory Imagery (Smell)

Appeals to the sense of smell.

1. *"The sharp tang of oranges and spice in the winter market"*
2. *"The scent of old books and dust in the library"*

3. *"Pine needles and earth after the rain"*
4. *"The acrid stench of smoke clung to his coat"*
5. *"The perfume of roses, cloying and thick in the summer heat"*
6. *"Fresh bread steaming in the morning air"*
7. *"A hint of lavender on her skin"*
8. *"The metallic smell of blood filled the air"*
9. *"Chlorine and sweat clung to the swimming pool"*
10. *"The musty odor of a forgotten attic"*

Gustatory Imagery (Taste)

Appeals to the sense of taste.

1. *"The tartness of lemon danced on her tongue"*
2. *"Chocolate melted bittersweet in his mouth"*
3. *"Salt from the sea lingered on his lips"*
4. *"The soup was a warm, savory promise"*
5. *"Her kiss tasted of red wine and secrets"*
6. *"The metallic tang of fear was bitter in his throat"*
7. *"Spices exploded on his palate—cumin, garlic, and fire"*
8. *"Frosted berries burst with sugary juice"*
9. *"The taste of burnt toast lingered, unwanted and sour"*
10. *"Mint chilled her mouth like winter in bloom"*

Kinesthetic Imagery (Movement/Tension)

Conveys movement or bodily sensations.

1. *"She floated across the floor like a leaf in wind"*
2. *"His heart pounded like a war drum"*

3. *"The dancers spun with wild abandon"*
4. *"Each footstep was a thud of determination"*
5. *"The horse galloped, muscles rippling under sunlit sweat"*
6. *"Adrenaline surged through her limbs like wildfire"*
7. *"He staggered as though the earth tilted beneath him"*
8. *"They lunged into the dark, breathless and frantic"*
9. *"The boat rocked gently with the ocean's sigh"*
10. *"The tightrope trembled beneath her trembling feet"*

Organic Imagery (Internal sensations or emotions)

Evokes internal experiences—hunger, fatigue, love, fear, etc.

1. *"A dull ache bloomed in her chest like grief taking root"*
2. *"His stomach churned with guilt and stale coffee"*
3. *"Loneliness wrapped around him like a winter coat"*
4. *"Joy bubbled in her like champagne in a glass"*
5. *"A knot of dread twisted in her gut"*
6. *"His limbs ached with the weight of despair"*
7. *"Peace settled on him like warm sunlight"*
8. *"Hunger gnawed at him with invisible teeth"*
9. *"Terror clutched his throat, choking speech"*
10. *"Love pulsed through her like a quiet heartbeat"*

Would you like this as a downloadable list or formatted for classroom or creative writing use?

You said:

can you add exotic examples?

ChatGPT said:

Absolutely — here's a **richly expanded list** of **exotic poetic imagery** samples, including rare and culturally diverse references. These examples are designed to evoke distant lands, unusual sensations, and vivid, otherworldly experiences.

Exotic Visual Imagery

1. *"Turquoise domes shimmered under a saffron sky"*
2. *"The desert bloomed with crimson hibiscus and golden fire lilies"*
3. *"Lanterns swayed on silk cords in the monsoon breeze"*
4. *"Peacocks strutted across the marble courtyard, tails trailing like jeweled fans"*
5. *"Moonlight silvered the temple roofs of Kyoto like frost on porcelain"*
6. *"Glaciers glinted like shattered glass beneath the aurora's green flame"*
7. *"Jade dragons coiled around the gate, their eyes set with onyx stones"*
8. *"The oasis flickered like a sapphire mirage in the ochre dust"*
9. *"The jungle canopy burned with parrots and sunlight"*
10. *"Caravans crawled across the dunes like ants over molten bronze"*

Exotic Auditory Imagery

1. *"The haunting call of a loon echoed across the fjord at dusk"*

2. *"Cicadas shrilled like electric threads in the Thai forest"*
3. *"Prayer calls tangled with birdsong over the old stone medina"*
4. *"Yak bells chimed softly in the Himalayan wind"*
5. *"Rain drummed on banana leaves like hollow conch shells"*
6. *"The bamboo grove rustled like a thousand whispers from another world"*
7. *"The didgeridoo's hum vibrated deep in the bones of the earth"*
8. *"Monks chanted, voices rising like smoke through incense"*
9. *"The market buzzed with bartering in six languages and the clink of spices"*
10. *"The crackle of cooking oil mingled with the sizzle of street firecrackers"*

Exotic Tactile Imagery

1. *"The silk of a sari brushed past her ankle like a dragonfly's wing"*
2. *"The humid air clung to her skin like warm syrup"*
3. *"Polished obsidian felt cold as starlight in her palm"*
4. *"His bare feet sank into volcanic sand still hot from the sun"*
5. *"The papaya's sticky juice slid between her fingers like honeyed lava"*
6. *"The camel's breath was damp and musky against his neck"*
7. *"The banyan tree's bark was gnarled and thick like elephant hide"*
8. *"A sudden wind carried the sting of jungle nettles on bare arms"*

9. *"The silk fan snapped open with a flick, like the bite of a cool breeze"*
10. *"Rain soaked her linen tunic, clinging like the skin of a mango"*

Exotic Olfactory Imagery

1. *"The heady perfume of frangipani hung in the island air"*
2. *"Jasmine and cardamom laced the night with memory"*
3. *"The jungle smelled of wet earth, rotting fruit, and ancient secrets"*
4. *"Incense wound through the temple—sandalwood, myrrh, and musk"*
5. *"The tang of pickled mango cut through the midday heat"*
6. *"Her skin bore the scent of coconut oil and sea salt"*
7. *"Spices danced on the breeze: turmeric, clove, star anise"*
8. *"The air was rich with the scent of ripe guava and smoldering charcoal"*
9. *"Camel sweat and old leather mingled with the dust of the Silk Road"*
10. *"A whisper of oud curled through the evening like a promise"*

Exotic Gustatory Imagery

1. *"The durian's flesh was creamy, pungent, like custard in a thunderstorm"*
2. *"The tang of tamarind burst sharp and sweet on her tongue"*

3. *"Lychee popped with perfumed nectar, bright as pearls in syrup"*
4. *"The tea was smoky and bitter, like fire and wind in a cup"*
5. *"Paprika, chili, and lime lit up her mouth like a fire festival"*
6. *"The mango tasted like sunshine steeped in velvet"*
7. *"Bitter cocoa and chili clashed on his tongue like lovers in heat"*
8. *"A sip of rosewater lingered like poetry on her lips"*
9. *"Dragonfruit fizzed with sweetness and strange familiarity"*
10. *"The soup carried the tang of lemongrass and secrets from the river"*

Exotic Kinesthetic Imagery

1. *"He moved like a panther across the red dirt, silent and watching"*
2. *"The elephant swayed, her gait hypnotic under the heavy sun"*
3. *"She danced barefoot in the rain, mud flying like petals"*
4. *"The dhow rocked lazily on the Persian Gulf, sails full of saffron wind"*
5. *"The falcon soared in tight spirals above the golden dunes"*
6. *"He leapt between rooftops in Marrakesh, nimble as a desert fox"*
7. *"They climbed the rice terrace steps like ants up a green staircase"*
8. *"The canoe glided over black water like a dream between stars"*
9. *"The sand shifted under each step as though the desert whispered 'go back'"*

10. *"She bent like a reed in the monsoon wind, and did not break"*

Would you like to compile this into a creative writing reference sheet or a printable guide for students or authors? I can format it for download.

You said:

can you make some sexy images?

ChatGPT said:

Certainly — here's a curated collection of **poetic imagery with sensual or sexy undertones**, crafted to be evocative, elegant, and suitable for literary or romantic writing. These examples balance suggestion and style, invoking intimacy through vivid language.

Sensual Visual Imagery

1. *"She slipped from the silk robe, moonlight tracing every curve like a lover's gaze"*
2. *"His shirt clung to him, damp with heat, revealing the quiet architecture of his body"*
3. *"The candlelight danced on her skin, gold and shadow playing over collarbones and hips"*
4. *"Lips stained with wine, her smile promised both danger and delight"*
5. *"The sheet fell from her shoulder like mist from the sea, baring sun-kissed skin"*

6. *"His eyes undressed her more gently than fingers ever could"*
7. *"She stood barefoot in the doorway, wearing only the dusk and his hunger"*
8. *"His breath caught as she stepped from the bath, dripping like a goddess from myth"*
9. *"Her glance lingered like fingertips down his spine"*
10. *"She wore red—not for fashion, but for fire"*

Sensual Tactile Imagery

1. *"His hand traced the hollow of her back like a secret being written in heat"*
2. *"Her skin was warm velvet beneath his fingers, pulsing with breath and want"*
3. *"Their bodies pressed together like verses longing to rhyme"*
4. *"The touch of her lips was soft as silk, fierce as flame"*
5. *"Goosebumps rose beneath his fingertips as he whispered her name into her collarbone"*
6. *"Her thighs tightened around him like a promise"*
7. *"His mouth left trails of warmth across her like fireflies in the dark"*
8. *"The heat of her breath melted into the hollow of his neck"*
9. *"His fingertips lingered at the waistband of her jeans, asking without asking"*
10. *"The sheets tangled around their legs like vines in bloom"*

Sensual Gustatory & Olfactory Imagery

1. *"Her kiss tasted of strawberries soaked in wine and moonlight"*
2. *"The scent of her skin—jasmine, sweat, and summer—made him forget to breathe"*
3. *"His mouth explored her like a ripe peach, sweet and dripping"*
4. *"Her breath carried the warmth of cinnamon and desire"*
5. *"They shared a pomegranate, juice staining their fingers and lips like sin"*
6. *"She smelled of salt and coconut, like the sea after a storm"*
7. *"The taste of her lingered on his tongue, warm and unspoken"*
8. *"He kissed the honey from her shoulder with reverence"*
9. *"The wine was rich and dark, like the curve of her laughter in his ear"*
10. *"She fed him fig and honey with her fingers, slow and deliberate"*

Sensual Auditory Imagery

1. *"Her sigh was a note only he could hear, soft as silk unraveling"*
2. *"The whisper of her name on his lips was the sound of longing made real"*
3. *"The rustle of her dress slipping to the floor made his breath hitch"*
4. *"Their laughter mingled like warm water and wine"*
5. *"Her moan rose like a prayer, hushed and holy"*
6. *"He spoke her name into the curve of her neck like a sacred vow"*
7. *"Each breath between them was a poem still being written"*

8. *"The low hum of pleasure escaped her lips, primal and beautiful"*
9. *"The rhythm of their heartbeats spoke louder than any words"*
10. *"Their silence was full of sound—skin against skin, breath against breath"*

Would you like these tailored for a specific tone—more romantic, more explicit, or perhaps for a particular setting (e.g., historical, fantasy, modern)? I can also format this into a PDF or a writing prompt generator.

You said:

how about unusual imagery?

ChatGPT said:

Absolutely — here's a collection of **unusual, striking, and unexpected poetic imagery** designed to break clichés, awaken the reader's senses in fresh ways, and ignite imaginative thinking. These examples lean surreal, metaphorical, and sometimes eerie or playful — excellent for poets, fiction writers, or lyricists aiming to stand out.

Unusual Visual Imagery

1. *"The sky peeled back like a bruised fruit, bleeding orange light"*
2. *"She wore silence like a glass dress—elegant, fragile, and about to shatter"*

3. *"His shadow dragged behind him like a tired animal"*
4. *"A neon halo flickered above her head, glitching like an old god"*
5. *"The trees leaned in, conspirators cloaked in green suspicion"*
6. *"Streetlights blinked like slow, indifferent eyes"*
7. *"The moon hung askew, like a button about to fall off the night"*
8. *"Her eyes were storms bottled in glass—beautiful and dangerous to hold"*
9. *"Clouds swam across the sky like thoughts trying to remember themselves"*
10. *"His grin cracked across his face like a fault line waiting to split"*

Unusual Auditory Imagery

1. *"The clock's ticking sounded like teeth chewing time"*
2. *"Her voice was velvet scraped over gravel—soft, but rough with memory"*
3. *"He laughed like a piano falling down a staircase—chaotic, brilliant, unrepeatable"*
4. *"Wind screeched through the alley like a violin out of tune with reality"*
5. *"The rain whispered confessions only the gutters understood"*
6. *"Each step echoed like punctuation in a sentence of dread"*
7. *"The radiator sighed like an old man remembering his first love"*
8. *"His voice buzzed in my ear like a fly that knew all my secrets"*
9. *"Her silence made a noise—like velvet being torn slowly"*

10. *"Thunder thumped like a drunken god falling down stairs"*

Unusual Tactile Imagery

1. *"The air was thick and slick, like breathing through honey left too long in the sun"*
2. *"His touch was cool and dry, like paper before it burns"*
3. *"The floor hummed under my feet, like it was hiding something alive"*
4. *"Her skin was textured like unfinished sentences"*
5. *"The cold bit like envy—sharp, invisible, hard to explain"*
6. *"He pressed against me like an unfinished thought"*
7. *"The fog curled around me like a lie you want to believe"*
8. *"Sweat dripped down his spine like guilt trying to escape"*
9. *"The sand scratched her legs like laughter with bad intentions"*
10. *"Touching him was like petting static—strange, electric, and risky"*

Unusual Olfactory Imagery

1. *"The air smelled of rain and raw ideas"*
2. *"Her perfume clung to the room like an argument left unresolved"*
3. *"The scent of burnt toast and old regret drifted from the kitchen"*
4. *"He smelled of leather, smoke, and the kind of night you don't tell stories about"*

5. *"The alley stank of damp cardboard and dreams too cheap to chase"*
6. *"A hint of ozone—like the sky had just been electrocuted"*
7. *"The perfume was floral, but tired, like a memory dressed up for a funeral"*
8. *"The theater smelled like popcorn and promises that wouldn't keep"*
9. *"The subway reeked of loneliness, metal, and wet wool"*
10. *"Her scent had gravity—it pulled memory and desire from places I didn't know I stored them"*

Unusual Gustatory Imagery

1. *"The soup tasted like nostalgia mixed with salt"*
2. *"His kiss tasted like lightning trapped in sugar"*
3. *"The cocktail fizzed on her tongue like champagne over a broken heart"*
4. *"The bread was warm, doughy, and as reassuring as a grandmother's hug through time"*
5. *"The coffee was bitter, like truth before you're ready for it"*
6. *"She bit into the fig and it bled sweetness and secrets"*
7. *"The chocolate melted like a slow confession"*
8. *"The ice cream tasted like childhood mixed with early disappointment"*
9. *"That wine had a backbone—it stood up in her mouth and made demands"*
10. *"It tasted like summer in the wrong month, confusing but welcome"*

Unusual Organic & Surreal Imagery

1. *"Grief grew inside her like moss on forgotten stone"*
2. *"His guilt was a wasp in his chest, always buzzing, never striking"*
3. *"Loneliness curled up behind her ribs and fell asleep"*
4. *"She carried joy like a cracked teacup—fragile and still useful"*
5. *"Hope itched under his skin like a fever you don't want to break"*
6. *"Love pooled in her throat, too thick to swallow, too dangerous to spit out"*
7. *"Anxiety crawled through his veins wearing boots"*
8. *"Desire fluttered behind her eyes like a moth desperate to escape the dark"*
9. *"Anger sat on his tongue like coins, metallic and ready to be thrown"*
10. *"Her soul knocked at her teeth, wanting out"*

Would you like a version of this tailored for fantasy, science fiction, horror, or a certain poetic style (e.g., surrealism, magical realism, absurdism)? I'd be glad to shape it further.

You said:

how about highly literary poetic imagery?

ChatGPT said:

Absolutely — here's a collection of **highly literary poetic imagery**, rich in allusion, layered metaphor, and elevated diction. These examples aim for **depth, beauty, and sophistication**, drawing on classical references, intellectual nuance, and emotional resonance — the kind you'd find in the works of poets like Rilke, Yeats, Plath, Eliot, or Louise Glück.

Visual Literary Imagery

1. *"The twilight peeled away from the earth like a palimpsest, revealing older shadows beneath"*
2. *"Her gown trailed behind her like an epic — each step a stanza, stitched in silk"*
3. *"He stood among the marble statues like an accusation carved from time itself"*
4. *"The orchard bloomed like a fevered memory, blossoms pale as unwritten prayers"*
5. *"Candlelight crawled across her face as if illumination were a hesitant visitor"*
6. *"The sky broke open like a psalm, bruised and golden at the edges"*
7. *"Columns of smoke curled like Sapphic verses, incomplete but yearning"*
8. *"The architecture of silence built itself between them like frost across a library window"*
9. *"The sea appeared not as water but as language unraveling — consonant foam, vowels of depth"*
10. *"He vanished into the crowd like myth—recognizable only by the ache he left behind"*

Philosophical/Abstract Literary Imagery

1. *"Desire unfolded within her like a theorem—precise, inevitable, terrifying in its elegance"*
2. *"Time passed like an elegy whose rhyme had long been forgotten"*
3. *"Grief nested in her throat like a bird that had mistaken her body for shelter"*

4. *"He wore detachment like a Renaissance mask—beautiful, sculpted, and meant to deceive"*
5. *"Loneliness bent around her like light around a dying star—visible, but unreachable"*
6. *"Hope felt like reading the final page of a book you dare not begin"*
7. *"The moment cracked open, revealing history coiled inside it like a serpent"*
8. *"He believed in love the way a blind man believes in constellations: abstractly, achingly"*
9. *"She spoke with the authority of ruins—weathered, eloquent, and irrefutable"*
10. *"The silence between them was not absence, but architecture"*

Auditory Literary Imagery

1. *"The wind spoke in a tongue older than grammar, braided with loss"*
2. *"Her voice fell through the room like Bach played underwater—distorted, but no less divine"*
3. *"The pages rustled like monastic robes shifting in an ancient cloister"*
4. *"Their argument echoed like cathedral bells — dignified in volume, violent in purpose"*
5. *"Rain tapped against the windowpane with the precision of a well-versed confession"*
6. *"His laughter cracked the air like a heresy whispered in a church"*
7. *"The cicadas screamed like souls caught mid-translation"*
8. *"His breath was music without notation—raw, instinctive, mortal"*
9. *"Even her silences were polyphonic"*

10. *"Words poured from him like ink from a divine hemorrhage"*

Tactile Literary Imagery

1. *"The book's leather binding felt like an ancestor's skin — worn, patient, familiar"*
2. *"The air hung heavy, like velvet soaked in ink and history"*
3. *"Her fingers moved across the piano like light touching broken glass—careful, luminous, doomed"*
4. *"The night pressed down like the weight of unread letters"*
5. *"Each step on the gravel path struck like punctuation in an unfinished poem"*
6. *"His hand brushed hers — the contact subtle as a footnote, and just as revelatory"*
7. *"Memory returned with the texture of wool—warm, coarse, and unwilling to let go"*
8. *"She wore her sadness like a second skin, tailored and invisible"*
9. *"The air brushed past like the hem of a ghost's garment"*
10. *"Her breath came in stanzas, fractured by restraint and rhythm"*

Olfactory & Gustatory Literary Imagery

1. *"The air smelled of cedarwood and forgotten things—like a philosophy abandoned mid-thought"*
2. *"The wine was a red so dark it tasted of myth—old gods and burial songs"*

3. *"Her skin held the scent of ink and rain — the smell of unwritten novels and stormlight"*
4. *"He tasted of salt and silence, like the sea after mourning"*
5. *"The bread's scent rose like a hymn — warm, homely, devout"*
6. *"The kitchen smelled of lavender and ash — comfort undercut by history"*
7. *"The orange's bitterness curled around her tongue like an unsent letter"*
8. *"The scent of burning leaves called forth childhoods she never lived"*
9. *"The soup had the warmth of a memory cooked slowly and served alone"*
10. *"His kiss tasted of smoke and syllables not yet spoken"*

Literary Romantic/Sensual Imagery (Subtle, Elevated)

1. *"She undressed like a sonnet unspooling — measured, rhythmic, and rich with hidden meaning"*
2. *"His hands knew her not as territory, but as text — something to be read, revered, reread"*
3. *"Their kiss was a question posed in the language of philosophers — impossible to answer cleanly"*
4. *"She lay beside him as if poetry had taken human form — exquisite, brief, unbearable"*
5. *"Their bodies folded into one another like parentheses in a secret verse"*
6. *"Desire hummed between them like an ancient harp played by trembling fingers"*
7. *"He touched her like an archivist — reverent, terrified to damage the parchment"*
8. *"Their sighs mingled like marginalia — intimate, telling, barely seen"*

9. *"She entered the room like a metaphor that demanded interpretation"*
10. *"In his arms, she felt not safe — but necessary"*

Would you like this formatted into a **PDF or printable writer's resource**? Or tailored to a specific genre like **modernist**, **romantic**, or **postcolonial literary style**? I'd be happy to extend it or refine it further.

You said:

how abouit imagery in the style of william black?

ChatGPT said:

Imagery in the style of **William Blake**—poet, painter, and mystic of the Romantic era—is rich with **visionary symbolism**, **spiritual intensity**, and **mythic duality**. His imagery blends **heaven and hell, innocence and experience, the divine and the carnal**, often expressed through vivid archetypes and stark contrasts.

Below is a collection of **original poetic imagery in the spirit of William Blake**, drawing from his mystical language, biblical allusions, and prophetic style.

Blakean Visual Imagery

1. *"I saw a lamb with stars in its eyes, and blood blooming like roses on its fleece"*

2. *"The sun unrolled itself like a scroll, spilling golden judgment upon the hills"*
3. *"A child of fire danced atop the tree of death, laughing in tongues of prophecy"*
4. *"Seraphim in chains sang hymns through gritted teeth of bone"*
5. *"The moon wept milk into the mouths of sleeping sinners"*
6. *"Iron trees bore fruit of flame in the gardens of lost virtue"*
7. *"Beasts with wings of parchment and ink flew from the breast of Wisdom"*
8. *"The face of God shone through the mirror, fractured into wrath and mercy"*
9. *"Stars fell like shattered eyes upon the trembling sea"*
10. *"A tiger licked the sky with tongues of bronze, eyes burning with the question of creation"*

Blakean Symbolic/Spiritual Imagery

1. *"My soul walked naked through the furnaces of desire, clothed only in innocence"*
2. *"The heart is a forge where angels and demons craft their war"*
3. *"I met Eternity beneath the tree whose roots reach into dream and death alike"*
4. *"The spirit is a mirror of fire—reflecting truth, but burning all who gaze too long"*
5. *"Sin sat beside me in the garden, wearing a crown of flowers and teeth"*
6. *"The mind is a temple where idols are carved from thought and fear"*
7. *"In every man sleeps a beast; in every beast, a divine spark longs to sing"*

8. *"Reason bound Imagination in chains, and called it order"*
9. *"From the womb of night was born a voice that sang of stars unborn"*
10. *"I held the sword of knowledge, and it whispered both salvation and ruin"*

Contrasting Dualities (Innocence & Experience)

1. *"The child smiles in the cradle, while Death rocks the other end"*
2. *"What is gentle in the lamb is terrible in the lion, yet both are love"*
3. *"Eden lies in shadow behind every city of light"*
4. *"The hand that heals may also bear the nails"*
5. *"Joy walked with a bloodied veil, for she had danced with sorrow"*
6. *"The tyrant prayed with lips of gold; the beggar cursed with lips of truth"*
7. *"Light may blind as surely as darkness may guide"*
8. *"He sang of peace, though his heart drummed with war"*
9. *"Every kiss of innocence leaves behind the breath of experience"*
10. *"Truth wore two faces: one of a dove, the other a flaming wheel"*

Animal & Elemental Imagery (Common in Blake)

1. *"A lion stood at the gates of dawn, roaring songs of judgment and mercy"*

2. *"The serpent curled around the tree, offering wisdom in the shape of a wound"*
3. *"A dove rose from ash, its wings inked with forgotten psalms"*
4. *"Thunder walked like a king through the valleys of men"*
5. *"The river sang of time, dragging bones and blossoms in equal measure"*
6. *"Wolves in priestly robes chanted prayers in the forest of illusions"*
7. *"The sky cracked open like a scroll, revealing the eyes of angels behind the stars"*
8. *"Flames whispered secrets to the stones of the temple"*
9. *"Rain fell like tears of the unborn, washing away the names of kings"*
10. *"A raven wore the crown of prophecy and pecked at the heart of man"*

Language Techniques Blake Used and That Appear Here

- **Archetypes**: lamb, tiger, child, serpent, angel, fire, star
- **Contrasts**: innocence vs experience, heaven vs hell, divine vs carnal
- **Visionary syntax**: archaic phrasing, biblical cadence, capitalized abstractions
- **Prophetic tone**: declarative, ecstatic, apocalyptic
- **Symbolism**: Light, fire, blood, tree, mirror, chains, song

Would you like to build a poem or prose piece in Blake's prophetic style using this imagery? I can help you generate an original poem *à la* Blake, or create a mythic framework similar to his **Songs of Innocence and of Experience** or **The Marriage of Heaven and Hell**.

You said:

how about in the style of walt whitman?

ChatGPT said:

Excellent choice. Walt Whitman's imagery is **expansive, democratic, bodily, cosmic, and deeply personal**—embracing the full sprawl of human experience and the natural world. His lines overflow with **catalogs**, **celebration**, **earthiness**, and **mystical union** between self and universe. Imagery in his style tends to be **sensory-rich, exuberant, and inclusive**—full of the **sacredness of the ordinary** and the **grandeur of the everyday**.

Here is a collection of **original poetic imagery in the style of Walt Whitman**:

Bodily & Earthly Imagery (Celebrating the Physical Self)

1. *"I sing the sweat of the worker's back, glistening beneath a merciful sun"*
2. *"The thighs of the swimmer gleam like seals slipping through salted water"*
3. *"The breath of the lover, warm and steady, is holy as scripture"*
4. *"O the tongue, pink banner of truth, tasting the meat of peaches and kisses alike!"*
5. *"The body is not shameful, not secret — it is the poem, the prophet, the nation"*

6. *"Belly rising and falling in sleep beside me — I adore it more than the moon"*
7. *"Fingers curled in soil, fingernails black with spring — that is prayer enough for me"*
8. *"Your spine is a river of lightning — I would walk barefoot across every vertebra"*
9. *"The musk of armpits in summer heat is the scent of real life, unedited"*
10. *"O hips of women and men, carriers of blood, of generations, of joy!"*

Nature & Democratic Imagery (The Shared World)

1. *"The blades of grass are messages from God — green tongues lapping at my ankles"*
2. *"The trees speak in thunder and leaf, witnesses to lovers and revolutions alike"*
3. *"I lie down with the stars, who do not question my worth"*
4. *"O the vast Mississippi — it carries the sweat of slaves, the dreams of boatmen, and still it sings"*
5. *"Even the fly on my hand belongs here — part of the great democratic dance of being"*
6. *"All the rivers know my name, and I know theirs — we were born from the same shout"*
7. *"I lean into the prairie wind as into a beloved's shoulder"*
8. *"I speak with the sunflowers, who are not shy to face the burning truth"*
9. *"The asphalt steaming after rain is a temple floor, and the passing bus my sermon"*
10. *"Every stone is sacred, every sidewalk a hymn, every gutter a verse in the great body of the world"*

Humanity & Collective Experience

1. *"The child playing stickball in Harlem is my brother, my youth, my joy"*
2. *"I eat with the mechanic, the widow, the thief, and the teacher — we bless the bread together"*
3. *"The soldier's torn boot, the mother's ache, the prisoner's sigh — I carry them all in my chest"*
4. *"I do not cast away the beggar or the banker — all are verses in the same unrhymed epic"*
5. *"The language of America is not only English — it is sweat, salsa, soot, and sunrise"*
6. *"I have slept in shacks and cathedrals — both felt like home"*
7. *"O the moan of the laborer and the moan of the lover are one sound to me"*
8. *"Whoever touches this page touches a man — hairy, hungry, holy, and laughing"*
9. *"I give myself to you, stranger, because we have shared the miracle of waking today"*
10. *"No one is excluded from the song I sing, not even the silent or the damned"*

Cosmic and Mystical Imagery (The Self as the Universe)

1. *"I am the comet streaking over Kansas, the quiver in your rib when you remember someone you lost"*
2. *"I float in the breath of the galaxy, and it smells of iron and lilacs"*
3. *"I am not one man — I am the wind that moves the wheat, the blood in your temples, the salt in your tears"*

4. *"O I have kissed the mouth of eternity and it tastes of grass and smoke and every woman I've loved"*
5. *"The atoms of Lincoln, the sweat of slaves, the milk of mothers — I carry it all in me"*
6. *"I died ten thousand times and was born ten thousand more, in bread lines and fields and hospital beds"*
7. *"I do not know God — I am God, and so are you, and so is the beetle crawling across my boot"*
8. *"The stars are freckles on the face of something I almost remember"*
9. *"My soul is the size of the world, and it fits under my hat"*
10. *"Time washes through me like a tide — and I remain, naked and singing"*

Key Whitmanian Traits in the Imagery Above

- **Free verse and repetition**
- **Celebration of the physical body and all its functions**
- **Use of catalogs/lists**
- **Spirituality rooted in the physical world**
- **Radical empathy and egalitarianism**
- **Sensory language grounded in real life**
- **Cosmic self-union — "I contain multitudes"**

Would you like help crafting an **original Whitman-style poem**, or perhaps a **writing prompt series** to explore Whitman's voice for your own themes? I can also help you mimic his cadence and line structure for your own material.

You said:

how about imagery in the style of t.s. elliot

ChatGPT said:

Certainly. Imagery in the style of **T.S. Eliot** is often:

- **Fragmented** and **allusive**, drawing from classical literature, religion, and mythology
- Marked by a **bleak urban modernism** — fog, soot, alleys, decay, disconnection
- Infused with **symbolic weight** (e.g., time, seasons, water, dust, light, silence)
- Both deeply personal and impersonally universal, moving from the **quotidian to the metaphysical**
- Frequently shifting between **the sacred and the absurd**, with flashes of despair, irony, and transcendence

Below is a set of original poetic imagery crafted in the **spirit of T.S. Eliot**, especially reminiscent of *The Waste Land*, *The Love Song of J. Alfred Prufrock*, and *Four Quartets*.

Urban & Modernist Imagery

1. *"The streetlamp sputtered like a dying mind, casting thoughtless shadows across the drizzle"*
2. *"Coffee cups stacked in towers, lipstick-rimmed and ghosted with breathless conversations"*
3. *"A woman in yellow gloves waited beneath the flickering 'EXIT' sign, mouthing words to no one"*
4. *"The train groaned through the fog like a sermon no longer believed"*
5. *"There is a silence in this city that smells of dust and unreturned letters"*

6. *"Shop windows blinked with indifferent light; mannequins watched with perfect stillness"*
7. *"Chimneys coughing soot into a sky that forgot how to be blue"*
8. *"Traffic lights clicked like rosary beads on an atheist's wrist"*
9. *"The stairwell smelled of boiled cabbage, regret, and dying orchids"*
10. *"Umbrellas clashed like black beetles at the hour of judgment"*

Psychological & Fragmented Imagery

1. *"Between one thought and the next, a small betrayal blooms"*
2. *"Voices echo in the vestibule of memory, mispronouncing the past"*
3. *"I saw myself in a window and did not wave"*
4. *"My thoughts peel like wallpaper in an empty parlor"*
5. *"The mirror reveals nothing but a list of undone mornings"*
6. *"We whisper to ourselves in the voices of strangers we've never met"*
7. *"Desire paces the hallway in shoes that don't quite fit"*
8. *"A cough, a bell, a song — all part of the same confession"*
9. *"Time folds back on itself like an unfinished letter returned unopened"*
10. *"I was not there, and yet I remember everything"*

Religious & Mythic Allusions

1. *"The angel passed over again, but this time did not knock"*
2. *"Three kings stood at the bus stop, staring into Styrofoam coffee cups"*
3. *"In the dust of the cathedral lay a postcard from Babylon"*
4. *"She anointed her wrists with perfume and uncertainty"*
5. *"The chalice is cracked, and the wine tastes of iodine"*
6. *"A fig tree withers in the corner of the room, unsummoned"*
7. *"The sea refused its dead, and the dead refused to return"*
8. *"Even the prophets turn their faces to the wall now"*
9. *"Bread was broken, but no one remembered the blessing"*
10. *"We walk past Golgotha on our way to buy cigarettes"*

Temporal & Wasteland Imagery

1. *"April came again, too early and too tired to matter"*
2. *"The seasons flicker like static behind the broken TV of memory"*
3. *"Even the clocks forget which way to tick"*
4. *"We bury the hours like bones, with ceremony and reluctance"*
5. *"Time pools in gutters with the last of the rain"*
6. *"Each year grows paler, like wine left out in the sun"*
7. *"What we called noon was only shadow"*
8. *"The harvest came, but no one remembered what to reap"*
9. *"Spring thawed nothing but silence"*
10. *"A minute's peace costs more than a year of talk"*

Symbolic & Philosophical Imagery

1. *"The soul hums behind the ribs like a radio between stations"*
2. *"We sip eternity in paper cups, distracted by sirens"*
3. *"What is faith but a whisper we pretend we didn't hear?"*
4. *"Truth flickers like moths against the cathedral glass"*
5. *"I carry the weight of unborn thoughts like coins in an old coat"*
6. *"Each word is a door left ajar in a house we forgot we built"*
7. *"God hides in the pause between heartbeats"*
8. *"We measure wisdom in footsteps walked away from ourselves"*
9. *"Hope wears a grey overcoat and watches from across the platform"*
10. *"I asked the question, and the echo answered in a different key"*

Techniques Reflected Here

- **Imagistic layering** (juxtaposition of the mundane and the symbolic)
- **Quotidian metaphysics** (the sacred embedded in the banal)
- **Ambiguous voice** (the "I" that is both personal and prophetic)
- **Historical and mythical dislocation**
- **Claustrophobic urbanity and sterile modern life**
- **Religious yearning amid spiritual erosion**

Would you like help crafting an *Eliot-style poem*, or working his tone into something you're writing? I can also help break down how to use **mythic structure**, **fragmentation**, or **objective correlative** in your own verse — all hallmarks of his technique.

You said:

how about imagery in the style of shelley?

ChatGPT said:

Absolutely. Percy Bysshe Shelley's poetic imagery is:

- **Lyrical and elevated**, often romantic and musical in cadence
- Rich in **natural symbolism**: wind, sky, fire, cloud, ocean, stars
- Marked by a sense of the **sublime** — the beautiful and the terrifying
- Full of **idealism and revolt**, infused with revolutionary or metaphysical overtones
- Often evoking **transience**: decay, transformation, death, and rebirth
- Overflowing with **mythic and elemental power**

His images often suggest that nature and imagination are forces of divine revelation or social awakening.

Here is a collection of **original poetic imagery in the style of Percy Bysshe Shelley**, drawing inspiration from works like *Ode to the West Wind*, *Prometheus Unbound*, and *Adonais*.

Elemental Imagery (Wind, Sky, Fire, Ocean)

1. *"The west wind howled like a prophet, scattering the dead thoughts of men like leaves"*
2. *"Clouds galloped across the firmament like wild mares, their manes aflame with dusk"*
3. *"The sea heaved like a heart that dreams of liberty"*
4. *"Lightning unbound its golden scrolls across the dome of heaven"*
5. *"The air burned with invisible fire, an exhalation of the soul of the world"*
6. *"Stars sang in silence, frozen music of the divine"*
7. *"The river broke its icy fetters and rushed singing toward the ocean's mouth"*
8. *"A single flame rose from the ashes of the pyre, whispering of immortality"*
9. *"The moon unveiled herself from mist, pale and chaste as forgotten sorrow"*
10. *"Snow fell like angelic feathers upon the shoulders of an unrepentant earth"*

Idealist and Revolutionary Imagery

1. *"The mind is a temple in ruins, yet its echo still sings of stars and justice"*
2. *"Truth, a burning lamp beneath the avalanche of time, flickers defiantly"*
3. *"Hope walks barefoot through the fire, untouched by ash or fear"*
4. *"The tyrant's throne, carved from centuries, trembled beneath the weight of song"*
5. *"Chains rust where ideas have flown"*
6. *"Imagination flared like lightning across the stagnant sky of empire"*

7. *"Even silence became rebellion when shaped by a poet's tongue"*
8. *"Words, those airy arrows, pierce the armor of the world"*
9. *"Liberty rose from the ruins, crowned not with gold, but with morning"*
10. *"The heart of man, though buried in dust, dreams still of flight"*

Natural and Lyrical Imagery

1. *"A violet trembled in the shadow of the rose, fragrant with hidden longing"*
2. *"Each petal held the sky in miniature, a fleeting firmament of color"*
3. *"The breeze played upon the lake like a harp touched by unseen fingers"*
4. *"A lark rose into heaven trailing songs like sunbeams"*
5. *"Dawn spread her limbs across the hills like a goddess waking from sleep"*
6. *"Rain fell like silver prayers on the thirsty fields"*
7. *"The forest sighed, a green cathedral full of unseen spirits"*
8. *"Evening painted the sky with resignation and soft defiance"*
9. *"The daisy looked up from the dirt, unafraid of heaven's judgment"*
10. *"The stars blinked like tears unshed by the infinite"*

Transience, Death, and Rebirth

1. *"Time devours its children, yet from their bones it grows flowers"*
2. *"Ashes drifted upward like thoughts freed from flesh"*
3. *"The grave was not silence, but the seedbed of song"*
4. *"Decay whispered to beauty, and beauty replied with fire"*
5. *"The dying leaves wrote elegies in gold upon the wind"*
6. *"The broken statue still dreamed of standing"*
7. *"He slept beneath the earth, but the grass hummed his name"*
8. *"Winter clasped the earth like a bridegroom of shadows, awaiting spring"*
9. *"From the death of the star was born a new cosmos"*
10. *"Memory, that eternal echo, preserved what flesh could not"*

Spiritual & Mythic Imagery

1. *"The soul is a bird of flame, longing for the azure beyond mortality"*
2. *"Zephyrs kissed the brow of Olympus and bore back dreams to men"*
3. *"Heaven's gates were not above but folded within the heart's last silence"*
4. *"Prometheus stood, chained but unbent, his fire leaking into the world"*
5. *"The spirit leapt like Icarus toward truth, unafraid of melting wings"*
6. *"Angels of dawn shook their golden locks and scattered wisdom across the world"*
7. *"The Muse wept fire and gave birth to a comet"*
8. *"Immortality sleeps in the cradle of every heartbeat"*
9. *"Eternity sings in the breath between one verse and the next"*

10. *"A single tear of Orpheus watered the withered rose of hope"*

Key Shelleyan Devices Reflected Here

- **Personification of nature** as active, divine, and revelatory
- Use of **wind, fire, stars, rivers, dawn, and sky** as vehicles of transcendence or revolution
- Elevated diction and **musicality of phrasing**
- **Idealist vision** blending political and metaphysical dimensions
- **Mythic figures** such as Prometheus, Icarus, angels, and muses as inner symbols

Would you like help composing a **Shelleyan ode**, or weaving this kind of Romantic imagery into your own poem or story? I can also help contrast Shelley's imagery with Byron's or Keats's if you're exploring the differences in Romantic voice.

You said:

how about imagery in the style of wordsworth?

ChatGPT said:

Beautiful direction. Imagery in the style of **William Wordsworth** is:

- Deeply **naturalistic** and reverent toward the quiet majesty of the natural world

- Rooted in **simplicity, memory, and emotional reflection**
- Focused on the **ordinary life** of rural people and the **sacredness of daily scenes**
- Often suffused with a tone of **solitude, innocence, and spiritual connection**
- Elevates **childhood, the senses, and recollection** as pathways to truth
- Frequently uses **soft, pastoral light** and the **gentle moods of weather and seasons**

Here is a set of original poetic imagery in the style of **William Wordsworth**, drawing from the spirit of *Tintern Abbey*, *The Prelude*, and *Ode: Intimations of Immortality*.

Pastoral & Natural Imagery

1. *"A single lamb stood bleating in the dew, soft as a prayer unspoken"*
2. *"The brook hummed its ancient song, threading moss and stone with silver patience"*
3. *"An old tree leaned like a shepherd at dusk, worn with wisdom, still and kind"*
4. *"The meadow rolled in golden hush, each stalk of wheat a quiet sun"*
5. *"Mist floated above the valley floor, veiling the world like a child's first thought"*
6. *"A lark rose, its cry vanishing into morning as if the sky were listening"*
7. *"Cows stood knee-deep in the river, as calm as clouds"*
8. *"A fern trembled under the weight of light, not wind"*
9. *"In the hush between raindrops, I heard the hill breathing"*

10. *"The moonlight lay on the pond like a mother's hand on a sleeping child"*

Childhood & Innocence Imagery

1. *"A boy ran laughing through the barley, his joy taller than the grain"*
2. *"The stone wall, crumbly and warm, held secrets only children know"*
3. *"Her bare feet stirred the earth with unknowing grace, as though she blessed it"*
4. *"We played where nettles grew, unafraid, and never stung"*
5. *"The sky seemed closer when I was small—its blue more mine"*
6. *"Each tree was a friend, each stream a song, and the wind a secret messenger"*
7. *"I drank from the spring with both hands, tasting the morning as if it would never end"*
8. *"The path we walked as children still remembers us, though we do not remember it"*
9. *"In her gaze lived a gentleness too wide for language"*
10. *"The innocence of a flower was once enough to stop me walking"*

Memory, Reflection, and Spiritual Quiet

1. *"The quiet of the evening wrapped itself around my thoughts like ivy"*
2. *"Though the hills have not changed, my heart approaches them more slowly now"*

3. *"Even the wind seems to recall how I once knelt among these stones"*
4. *"A robin's call broke the silence like memory entering the mind"*
5. *"The lake receives my face and gives it back softened by stillness"*
6. *"This tree I once climbed now shelters me in shadow and thought"*
7. *"The sun on this path feels like forgiveness"*
8. *"I spoke to no one, and yet the whole field answered"*
9. *"In solitude I find not absence, but presence more complete than any crowd"*
10. *"The grass remembers my footprints from a time before I had burdens"*

Spiritual & Sublime Imagery

1. *"There is divinity in the hush of bees around the foxglove"*
2. *"Heaven rests lightly on the brow of the hill"*
3. *"The daffodils nod not for beauty's sake, but because they know joy"*
4. *"A single beam of sun on the lake opens eternity in my chest"*
5. *"The soul moves gently through the world when it listens"*
6. *"I have knelt at the altar of a fallen leaf"*
7. *"The holiest light is the one that makes no sound"*
8. *"A field in spring is scripture enough"*
9. *"In the stillness between two birdsong notes, I felt the sacred breath of being"*
10. *"To walk alone among the hills is to be accompanied by the infinite"*

Traits of Wordsworth's Imagery in These Examples

- Use of **plain, lucid language** elevated by musical phrasing
- Reverence for the **natural world as a moral and emotional teacher**
- Emphasis on **quiet moments** and **subtle observation**
- Nature as a mirror for **memory, personal growth, and spiritual truth**
- **Children and the past** as vessels of purity and forgotten understanding

Would you like help composing a **Wordsworthian lyric or reflective nature poem**? I can also help pair his imagery style with journal entries, pastoral prose, or meditative essays if you're writing in mixed forms.

You said:

how aBOUT romantic imagery?

ChatGPT said:

Got it! **Romantic imagery** broadly refers to the style characteristic of the Romantic movement (late 18th–early 19th century), but it's often associated with a few key qualities:

- **Intense emotion and individual experience** — passion, longing, melancholy, awe
- **Nature as a living, spiritual force** — vast landscapes, storms, wilderness, night skies

- **The sublime** — feelings of awe mixed with terror or wonder, confronting infinity or death
- **Imagination and dreams** — visions, myth, fantasy, the mysterious
- **Rebellion and freedom** — escape from societal constraints, yearning for transcendence
- **Exotic, historical, or medieval elements** — castles, ruins, distant lands, legends

Romantic imagery blends the physical and emotional worlds in vivid, often dramatic ways.

Here's a rich **list of original poetic imagery** in a Romantic vein, synthesizing these themes:

Nature as Sublime and Spirit

1. *"The mountains loomed like ancient gods, their crowns lost in storm-wrapped clouds"*
2. *"The ocean's breath was a fierce hymn, roaring freedom into the shattered cliffs"*
3. *"A lightning bolt cleaved the night's black veil, a sudden godlike revelation"*
4. *"The wildflower tangled with the thorn like a lover's desperate embrace"*
5. *"The moon hung, swollen and red, as if bleeding secrets to the dark woods"*
6. *"Fog slithered over the moor, a ghostly veil drawn tight across the land's wounds"*
7. *"A river sang with ancient sorrow, carving grief into the valley's bones"*

8. *"The forest breathed deep and slow, its heart beating in shadow and light"*
9. *"Stars burst like shattered glass in the midnight sky, sparks of eternity"*
10. *"Dawn spilled like molten gold, setting fire to the sleeping earth"*

Passion and Emotional Intensity

1. *"Her eyes were storms that dared the sky to break"*
2. *"Love's flame licked the edges of my reason, burning all caution away"*
3. *"The heart aches like a caged bird desperate for the wild wind"*
4. *"His voice was velvet thunder, shaking my soul awake"*
5. *"We danced beneath a sky bruised purple with longing"*
6. *"The touch of her hand was a secret poem written on my skin"*
7. *"Desire poured through me like a river swollen with spring rain"*
8. *"I held her absence like a fragile bloom, trembling yet alive"*
9. *"Passion's fire burned with the cold fury of a winter storm"*
10. *"The kiss tasted of rebellion and the sweetness of forbidden fruit"*

Imagination, Dreams, and Mystery

1. *"In dreams, the moon spoke to me in riddles woven from silver threads"*

2. *"A phantom castle rose from the mist, built of whispers and forgotten prayers"*
3. *"The wind carried songs of lands unseen, where time loops in endless twilight"*
4. *"Her shadow danced with the flicker of candlelight, a wraith of memory and hope"*
5. *"Stars blinked secrets into the void, their language a riddle no ear could catch"*
6. *"The night unfurled its velvet wings, folding me into the abyss of wonder"*
7. *"Clouds sailed like ghost ships across a sea of dreamless sky"*
8. *"The enchanted garden hummed with a magic older than any tongue"*
9. *"A raven's cry split the silence, a dark note in the symphony of night"*
10. *"The labyrinth of my mind twisted with echoes of forgotten legends"*

Exotic, Historical, and Mythic

1. *"Ruins rose like the bones of forgotten gods, half-swallowed by ivy and time"*
2. *"A warrior's sword gleamed beneath the blood-red moon, hungry for glory and fate"*
3. *"The minaret's shadow stretched long across the desert, a sentinel of ancient dreams"*
4. *"Silks shimmered in the market, a tapestry of distant suns and secrets"*
5. *"The harp's song wove spells that stirred the ghosts beneath the ruined keep"*
6. *"Chariots thundered across the horizon of memory, echoes of a lost empire"*

7. *"The siren's voice rose like a curse from the depths, promising oblivion and ecstasy"*
8. *"The enchanted forest whispered tales of knights and enchanted maidens"*
9. *"A crystal goblet overflowed with wine as red as the blood of forgotten kings"*
10. *"The tapestry depicted battles where gods and mortals danced the dance of fate"*

Freedom, Rebellion, and Transcendence

1. *"Chains shattered like glass beneath the feet of the dreaming rebel"*
2. *"The eagle soared beyond the prison of earth, wings stretched toward the infinite"*
3. *"Firestorms of change raged behind the eyes of the outcast poet"*
4. *"We broke the silence with shouts that carved new worlds from the ashes"*
5. *"The soul leapt like a flame from the ruins of despair"*
6. *"The dawn exploded like a riot, setting the sky ablaze with possibility"*
7. *"Stars aligned like secret compasses guiding lost wanderers home"*
8. *"The river refused to be tamed, singing its wild song through every canyon"*
9. *"He danced on the edge of chaos, fearless as a storm"*
10. *"The night cracked open, revealing the infinite fire beneath"*

If you want, I can help you **write a full Romantic poem** or create a **Romantic-themed scene or story** using this kind of vivid, passionate imagery. Just say the word!

You said:

how about symbolist imagery?

ChatGPT said:

Absolutely! **Symbolist imagery** is fascinating—very different in style and mood from Romanticism or Wordsworthian naturalism.

What defines Symbolist imagery?

- **Highly suggestive, indirect, and evocative** rather than explicit
- Often **dreamlike, mysterious, and ethereal**
- Uses **symbols and metaphors that point beyond the visible**, aiming to evoke emotions and states of mind rather than concrete realities
- Prefers **musicality, color, and mood** over clear narrative or description
- Frequently draws on **myth, mysticism, the occult, and esoteric themes**
- Plays with **shadows, mirrors, veils, twilight, and ephemeral light**
- Embraces the **inner life, subconscious, and spiritual realms**

Symbolist poets include: Stéphane Mallarmé, Charles Baudelaire, Paul Verlaine, Arthur Rimbaud, and later W.B. Yeats and Rainer Maria Rilke.

Here's a set of original poetic imagery inspired by Symbolist aesthetics:

Dreamlike & Ethereal Imagery

1. *"Veils of opal mist drifted between the silent towers of memory"*
2. *"A silver harp hung from the twilight sky, plucked by unseen fingers"*
3. *"The moon's pale smile dissolved into a pool of whispering shadows"*
4. *"Lanterns of smoke floated like ghosts in the garden of forgotten dreams"*
5. *"Stars bled indigo tears into the night's velvet shroud"*
6. *"The river murmured secrets in a tongue older than sleep"*
7. *"A ghostly rose unfolded beneath the breath of a languid wind"*
8. *"The cathedral of dusk echoed with the sighs of vanished saints"*
9. *"Mirrors shattered in slow motion, fracturing time like stained glass"*
10. *"A faint pulse throbbed beneath the veiled surface of the silent lake"*

Symbols of the Inner and Spiritual

1. *"The eye of the soul opened, a dark pearl in the ocean of night"*
2. *"A black flame flickered in the hollow of forgotten desires"*
3. *"The labyrinth of dreams wound endlessly beneath the skin"*
4. *"Silver serpents of thought slithered through the garden of silence"*
5. *"The heart was a crystal prism, fracturing sorrow into scattered light"*
6. *"Veins of starlight traced the map of invisible longings"*
7. *"The shadow whispered riddles to the restless moon"*
8. *"A solitary figure drifted like a smoke-wraith through corridors of memory"*
9. *"The soul's chalice brimmed with the nectar of forgotten gods"*
10. *"Time's fingers wove a tapestry of silence and shattered dreams"*

Colors, Sounds, and Musicality

1. *"Lavender sighs wove through the orchard's silver twilight"*
2. *"The bell tolled in waves of violet and gold, echoing in the marrow of night"*
3. *"Amber echoes spilled from the cracked urn of a dying star"*
4. *"Crimson petals drifted on the breath of a distant, forgotten song"*
5. *"The night was a symphony of blue shadows and murmured prayers"*

6. *"The garden hummed with a chorus of sapphire-winged dreams"*
7. *"Cobalt waves crashed in silent rhythm against the shore of sleep"*
8. *"Gossamer threads of sound unraveled in the still air like silver smoke"*
9. *"A violet flame flickered in the heart of the labyrinth"*
10. *"The dawn's first light was a whisper of molten gold and forgotten sighs"*

Mysticism, Occult & Esoteric

1. *"The alchemist's star burned cold and distant behind a veil of smoke"*
2. *"Runes glimmered faintly beneath the dust of ancient tomes"*
3. *"The silent oracle's eyes held galaxies folded in shadow"*
4. *"A serpent coiled in the circle of forgotten rites"*
5. *"Candles flickered like dying hopes in the sanctuary of silence"*
6. *"The midnight chalice overflowed with liquid moonlight and despair"*
7. *"Phantoms danced in the crypt of forgotten promises"*
8. *"A black rose bloomed in the garden where no sun dared linger"*
9. *"The secret language of stars whispered through the veils of night"*
10. *"The shadow priestess traced symbols in the dust, summoning the unseen"*

Symbolist Imagery Traits Here:

- Use of **ambiguity and layered metaphor** invites multiple interpretations
- **Sensory fusion** — colors, sounds, and smells blend into a synesthetic experience
- Imagery aims to **evoke moods** rather than describe objects
- Frequent use of **contrasting light and dark, presence and absence, surface and depth**
- A **focus on the mystical and unseen realms** beyond everyday reality

If you want, I can help you write a **symbolist-style poem or prose**, or show how to blend this style into modern poetry or visual art descriptions. Would you like that?

You said:

how about feminist imagery?

ChatGPT said:

Absolutely! **Feminist imagery** in poetry and literature often centers on:

- **Empowerment, resilience, and agency** of women
- Challenging traditional gender roles and stereotypes
- Celebrating the **female body, strength, and sexuality** unapologetically
- Exploring **solidarity, sisterhood, and shared experience**
- Reclaiming or reinterpreting symbols historically used to oppress women

- Highlighting themes of **liberation, self-discovery, and voice**
- Often grounded in **both the personal and political**

Feminist imagery can be bold, tender, fierce, or revolutionary — and it uses symbols drawn from nature, history, myth, and everyday life.

Here's a set of original poetic imagery inspired by feminist perspectives:

Nature as Feminine Power and Rebirth

1. *"Roots knot deep in earth, unyielding, weaving strength beneath the surface"*
2. *"The river carries stories women whispered for centuries, relentless and clear"*
3. *"A wild rose blooms through cracked stone, unafraid of the weight of silence"*
4. *"The moon's fierce glow cradles night like a mother guarding dreams"*
5. *"She rises like dawn, fierce as fire, soft as dew on stubborn grass"*
6. *"Branches stretch wide, like open arms in welcome and defiance"*
7. *"Her breath is wind that bends but does not break the tallest pine"*
8. *"Seeds of revolution germinate in the soil of quiet courage"*
9. *"The ocean's tide is her pulse—strong, steady, unstoppable"*

10. *"Vines climb walls of history, reclaiming space once denied"*

Strength, Resilience, and Agency

1. *"Fists rise like storm clouds, dark with promise and thunder"*
2. *"She wears scars like medals, each one a story of survival and victory"*
3. *"The armor she forges is made of laughter, tears, and relentless hope"*
4. *"Her voice breaks chains forged in silence"*
5. *"Eyes blazing with unspent fire, she moves through shadows unshaken"*
6. *"The heartbeat of resistance drums beneath her skin"*
7. *"Her steps carve paths in wild forests where no map dared lead"*
8. *"Like a phoenix she rises from ash, crowned in self-made light"*
9. *"She is the architect of her own story, building with hands fierce and tender"*
10. *"The quiet in her gaze is the calm before the revolution"*

Celebration of the Female Body and Sexuality

1. *"Curves like rivers, flowing with the grace of ancient landscapes"*
2. *"Her skin is a canvas painted by moonlight and memory"*
3. *"Desire blooms within her like wildflowers after rain—untamed and true"*

4. *“Fingers trace the map of her own becoming, fearless and intimate”*
5. *“Her laughter ripples like silk, soft but impossible to contain”*
6. *“The scent of jasmine lingers where her shadow falls”*
7. *“Eyes that hold storms and sunsets in equal measure”*
8. *“She wears her sensuality like a cloak woven from starlight”*
9. *“The rhythm of her body beats a song of freedom and joy”*
10. *“Her lips, a red comet burning through the night’s restraint”*

Sisterhood, Solidarity, and Shared Experience

1. *“Hands intertwined like braided rivers, unstoppable in their flow”*
2. *“Voices rising in chorus, a tide that breaks walls and mends hearts”*
3. *“Together, they are a forest, roots entwined beneath the surface”*
4. *“Eyes meet across the room, a silent pact of shared strength”*
5. *“Their laughter threads a tapestry woven from resilience and hope”*
6. *“The stories they tell echo like ancient songs, fierce and enduring”*
7. *“A circle of women glowing in the firelight, unbroken and eternal”*
8. *“They build bridges with hands calloused by both labor and love”*
9. *“In their embrace, the world softens, bends, and begins anew”*

10. *"Sisterhood blooms wild and free, a garden no storm can wither"*

Liberation, Voice, and Revolution

1. *"Her scream splits the night, shattering centuries of silence"*
2. *"Words flow like rivers breaking dams, unstoppable and pure"*
3. *"She stands on the edge of dawn, unafraid to leap into new worlds"*
4. *"Chains crumble beneath the weight of her untamed song"*
5. *"The pen in her hand is a sword forged in the fires of truth"*
6. *"She paints revolution in strokes bold and unyielding"*
7. *"Her shadow stretches long over the ruins of old orders"*
8. *"Each step forward is a dance of defiance and hope"*
9. *"The fire in her eyes is the dawn of all new beginnings"*
10. *"From silence, she builds a roar that shakes the foundations"*

If you want, I can help you craft a **feminist poem, manifesto, or prose** using these images — or explore how feminist imagery appears in literature from different cultures and periods. Would you like to try writing something together?

You said:

how about freudiam imagery?

ChatGPT said:

Absolutely! **Freudian imagery** in poetry and literature draws on the themes and symbols from psychoanalysis, especially those introduced by Sigmund Freud. This kind of imagery explores the **unconscious mind**, **repressed desires**, **dreams**, **sexuality**, **anxiety**, **the id, ego, and superego**, and often the tension between **civilization and primal instincts**.

Key features of Freudian imagery:

- **Dreamlike, surreal, and symbolic** — much like the content of dreams and unconscious thoughts
- Uses **archetypal symbols** (e.g., water as the unconscious, caves or tunnels as the womb or unconscious mind, mirrors for self-awareness, masks, shadows)
- Explores **sexuality, repression, and forbidden desires** often in veiled or symbolic ways
- Focuses on **conflicts within the psyche**—fear, guilt, aggression, desire, identity
- Often evokes **uncanny, eerie, or unsettling atmospheres**
- Includes motifs of **birth, death, decay, and transformation**

Here's a set of original poetic imagery inspired by Freudian themes:

Dream and Unconscious Imagery

1. *"A labyrinth of mirrors reflecting fractured selves in endless corridors"*
2. *"Beneath the skin, a river of shadows flows—dark, silent, and unstoppable"*
3. *"Dreams drip like melting wax from the candle of consciousness"*
4. *"A buried garden grows with forbidden fruit, roots tangled in silence"*
5. *"Whispers crawl from the depths like serpents coiled in the night"*
6. *"The mind's eye watches from behind a veil of shifting smoke"*
7. *"A locked chest pulses with the heartbeat of forgotten fears"*
8. *"The shadow self lurks beneath the mask of day"*
9. *"Echoes of laughter twist into screams inside the cavern of memory"*
10. *"A silent scream floats beneath the surface of sleep, unseen but felt"*

Sexuality and Desire

1. *"Flames lick the walls of a dark chamber, burning with unspoken hunger"*
2. *"Her body a map of secret rivers flowing beneath fragile skin"*
3. *"The touch lingers like a ghost, stirring embers in the depths"*
4. *"Lips press like shadows, tasting the edge of forbidden realms"*
5. *"Desire coils like a serpent in the garden of the mind"*

6. *"Veins throb with the pulse of a silent, relentless yearning"*
7. *"The scent of jasmine mingles with the ache of buried secrets"*
8. *"Fingers trace the silhouette of dreams on trembling flesh"*
9. *"A crimson tide surges beneath the calm surface of restraint"*
10. *"The night hums with the vibration of unspeakable longing"*

Conflict, Anxiety, and the Psyche

1. *"Chains rattle in the darkness of a mind torn by silent wars"*
2. *"The gaze of the superego cuts like a blade through veils of desire"*
3. *"Fear coils tight, a black snake tightening around the throat"*
4. *"Masks crack under the pressure of hidden truths"*
5. *"The ego balances on a knife's edge between chaos and order"*
6. *"Silent screams echo in the corridors of a restless mind"*
7. *"A house of glass shatters beneath the weight of unspoken guilt"*
8. *"The shadow creeps closer, swallowing light and hope alike"*
9. *"Thoughts fracture like broken ice on a frozen lake"*
10. *"The heart is a dark well, filled with the echoes of suppressed fears"*

Birth, Death, and Transformation

1. *"A cocoon cracks open, spilling shadows into uncertain dawn"*
2. *"Roots twist deep beneath the earth, seeking the light of rebirth"*
3. *"Ashes scatter on the wind, carrying the ghosts of old selves"*
4. *"The chrysalis trembles in the silence before flight"*
5. *"Bones gleam like forgotten promises beneath the soil of memory"*
6. *"The night folds over the world like a velvet shroud"*
7. *"A phoenix rises from the ruin of dreams, feathers dark with fire"*
8. *"Death whispers like a lover in the dark, tender and inevitable"*
9. *"Veins pulse with the rhythm of endings and beginnings entwined"*
10. *"The mirror reflects not the face, but the shadow behind it"*

Freudian Imagery Traits:

- **Dense symbolism and metaphor** that taps into deep psychological layers
- Frequent use of **dark, shadowy, and fluid images** to suggest unconscious processes
- Themes of **sexuality, repression, internal conflict, and transformation**
- A sense of **mystery, unease, or uncanny strangeness**

If you want, I can help craft a **Freudian-themed poem or short story** full of these images, or analyze a piece of writing with Freudian imagery in mind. Would you like to write something together?

You said:

how about jungian imagery?

ChatGPT said:

Absolutely! **Jungian imagery** is deeply rooted in the concepts introduced by Carl Gustav Jung, focusing on the **collective unconscious**, **archetypes**, **myth**, and the **process of individuation**—the journey toward self-realization and wholeness.

Key characteristics of Jungian imagery:

- **Archetypes** like the Shadow, Anima/Animus, the Self, the Hero, the Wise Old Man/Woman, the Trickster
- Symbols of **wholeness, duality, transformation, and integration**
- Use of **mythological, natural, and cosmic symbols** as representations of psychic forces
- Imagery that reflects **inner journeys**, encounters with the unconscious, and moments of revelation
- Frequent motifs: **mirrors, labyrinths, mandalas, the tree of life, the journey/path, dark waters, light emerging from darkness**
- Emphasis on **balance between opposites** (light/dark, conscious/unconscious, masculine/feminine)
- Imagery that evokes the **numinous**—the mysterious, sacred, and awe-inspiring

Here's a set of original poetic imagery inspired by Jungian symbolism:

Archetypes and Inner Figures

1. *"The shadow stretches long beneath the surface, a silent twin in the depths"*
2. *"An anima's voice sings through the forest, soft as moonlight on ancient stones"*
3. *"The wise woman's eyes hold the stars, deep wells of forgotten knowledge"*
4. *"A trickster dances on the edge of night, weaving riddles in silver thread"*
5. *"The hero's path winds through fire and shadow, marked by scars of courage"*
6. *"The self blooms like a mandala, petals unfolding in endless symmetry"*
7. *"A father's silhouette looms at the horizon, a mountain both stern and guiding"*
8. *"The mother earth cradles the lost child in roots and soil"*
9. *"The twin flames flicker—dark and light—dancing in eternal embrace"*
10. *"A serpent coils around the tree of life, guarding the gate to transformation"*

Symbols of Wholeness and Transformation

1. *"The labyrinth's walls whisper secrets to those who dare to wander within"*
2. *"Mirrors shatter and reassemble, reflecting fragments of the true self"*
3. *"A phoenix rises from ashes braided with threads of gold and shadow"*
4. *"Dark waters mirror a sky shattered by dawn's first light"*
5. *"The tree of life stretches its limbs toward the cosmos, roots entwined in earth's heart"*
6. *"A circle of stones glows softly beneath the watchful eye of the moon"*
7. *"The spiral winds inward, folding time into the fold of eternity"*
8. *"A chalice brims with liquid starlight, the nectar of rebirth"*
9. *"The sun and moon entwined in a silent dance, weaving day into night"*
10. *"A bridge of light arches over the abyss, a path from shadow to self"*

Inner Journeys and Mystical Encounters

1. *"Beneath the surface, the unconscious hums with the songs of forgotten ancestors"*
2. *"A silent river carries the weight of unspoken truths and hidden fears"*
3. *"The cave's mouth yawns wide, an invitation to the soul's descent"*
4. *"Stars glimmer like ancient runes, guiding the seeker through the night"*
5. *"A solitary figure stands at the crossroads, where worlds fold and unfold"*

6. *"The veil between waking and dreaming flutters like a moth's fragile wing"*
7. *"Lanterns drift over dark waters, flickering with hope and revelation"*
8. *"The whisper of the forest holds the memory of a thousand lifetimes"*
9. *"A serpent sheds its skin beneath the watchful gaze of the sacred tree"*
10. *"The journey spirals deeper, into the heart of the unknown"*

Numinous and Sacred Symbols

1. *"A mandala blooms in the center of the soul, infinite and whole"*
2. *"Sacred geometry pulses beneath the surface of waking life"*
3. *"The sacred flame burns steady in the temple of the heart"*
4. *"A star falls, scattering seeds of light across the dark canvas of night"*
5. *"The crossroads gleam with the shimmer of possibility and fate"*
6. *"An ancient rune glows faintly on the traveler's palm, a promise and a curse"*
7. *"The sacred dance of opposites weaves the fabric of reality"*
8. *"The eclipse cloaks the world in mystery, revealing hidden depths"*
9. *"A chalice carved from crystal captures the light of the infinite"*
10. *"The silence between heartbeats holds the secret of the cosmos"*

Jungian Imagery Traits:

- Rich, archetypal symbolism tapping into universal psychic patterns
- A balance of **light and dark, conscious and unconscious, masculine and feminine**
- Emphasis on **spiritual and psychological wholeness and transformation**
- Visual metaphors for the **inner journey and integration of the self**
- Mystical, cosmic, and mythic tones with a timeless feel

If you'd like, I can help you **write a Jungian-inspired poem or story**, or guide you through interpreting a text using Jungian symbolic imagery. Would you like to explore that?

You said:

how about japanese style imagery?

ChatGPT said:

Absolutely! **Japanese-style imagery** in poetry—especially in traditional forms like **haiku**, **tanka**, and **waka**—is renowned for its:

- **Simplicity, subtlety, and minimalism**
- Deep connection to **nature and the seasons (kigo)**
- Use of **suggestion and implication** rather than explicit description

- **Ephemeral beauty**—highlighting fleeting moments (mono no aware)
- Juxtaposition of images to evoke mood and reflection
- Often incorporates **natural phenomena, flora, fauna, and landscapes** with symbolic meaning
- Emphasis on **quiet contemplation, stillness, and transience**

Some typical motifs in Japanese poetic imagery:

- Cherry blossoms (sakura) as symbols of fleeting beauty and mortality
- Autumn leaves (momiji) signaling change and impermanence
- The moon as a symbol of loneliness, beauty, or enlightenment
- Mist, dew, and snow as evocations of delicate transience
- Bamboo, plum blossoms, and pine as symbols of resilience and endurance
- Seasonal references that evoke emotional states
- Birds (cuckoo, crane, nightingale) and insects (cicadas) to mark time and mood

Here's a set of original Japanese-style poetic imagery inspired by those traditions:

Seasonal & Nature Imagery

1. *"Cherry petals drift softly—brief whispers of spring's farewell"*
2. *"The cicada's song fades—summer's heat thins like morning mist"*
3. *"A single maple leaf twirls in the chill breeze, autumn's breath"*
4. *"Frost clings to pine needles, silent guardians of winter's hush"*
5. *"Plum blossoms awaken beneath pale clouds, promise in each bud"*
6. *"The moonlight slips through bamboo stalks, silver threads on dark silk"*
7. *"Dew gathers like pearls on spiderwebs, fragile jewels at dawn"*
8. *"Crickets chirp beneath the red maple, a lullaby for fading light"*
9. *"Snow melts quietly, unveiling earth's soft, muted palette"*
10. *"A heron stands still by the quiet pond, waiting for the perfect moment"*

Ephemeral Beauty & Emotion

1. *"A fleeting shadow flutters past—like a sigh caught on the wind"*
2. *"The scent of wisteria lingers, a ghost of summer's embrace"*
3. *"Moonlight brushes the surface of the lake—silver and serene"*
4. *"Mist rolls over the hills, blurring the boundary between dream and dawn"*
5. *"A spider's web glistens with morning dew—a fragile world revealed"*

6. *"Soft rain taps the temple roof—a meditation on passing time"*
7. *"The echo of a distant bell, fading into twilight's gentle arms"*
8. *"Clouds gather like whispered secrets above the silent garden"*
9. *"The empty swing sways slowly—a memory of laughter long gone"*
10. *"A plum tree bends beneath snow's weight, bowed but unbroken"*

Simplicity and Suggestion

1. *"Bamboo shadows stretch—quiet poems in shifting light"*
2. *"A single crow calls, breaking the stillness of the morning"*
3. *"Rice fields mirror the sky—patches of blue held in earth's hand"*
4. *"A lantern flickers—tiny stars caught in glass"*
5. *"Fallen leaves rest on a stone path—maps of forgotten journeys"*
6. *"The distant peak wears a crown of cloud—silent and aloof"*
7. *"Fishermen's boats drift home, floating like ink strokes on water"*
8. *"A butterfly pauses—breath held in a moment's hush"*
9. *"The wind stirs the grass—soft whispers of an unseen guest"*
10. *"A tea bowl warms slowly in hands—stillness made tangible"*

Japanese poetic imagery traits:

- Delicate, restrained, often **using nature to mirror human emotions**
- Deep appreciation of **transience and impermanence (mono no aware)**
- Emphasis on **evoking a mood with very few words or images**
- **Seasonal references** that root poems in time and place, connecting the personal to nature's cycles
- Often **quiet, contemplative, and meditative**

Would you like me to help craft a **haiku or tanka** using this kind of imagery? Or explore a modern poem inspired by Japanese aesthetics?

You said:

how about mythological imagery?

ChatGPT said:

Certainly! **Mythological imagery** in literature and art draws upon the rich tapestry of myths, legends, and folklore to convey deeper meanings, explore universal themes, and evoke powerful emotions. These symbols and motifs often transcend cultural boundaries, resonating with audiences across time and space.

Key Characteristics of Mythological Imagery

- **Archetypal Symbols**: Central figures like gods, heroes, monsters, and mythical creatures that embody universal human experiences and emotions.
- **Epic Narratives**: Stories of creation, destruction, transformation, and moral lessons that reflect the complexities of life and the cosmos.
- **Moral and Philosophical Themes**: Exploration of concepts such as hubris, fate, sacrifice, and the battle between good and evil.
- **Cultural Reflection**: Mirrors the values, fears, and aspirations of the societies that birthed them.

Examples of Mythological Imagery in Literature

1. **Prometheus**: In Greek mythology, Prometheus defied Zeus to bring fire to humanity, symbolizing rebellion, enlightenment, and the quest for knowledge. This myth has inspired countless works, including Mary Shelley's *Frankenstein*, where the protagonist's overreaching ambition mirrors Prometheus's defiance .
2. **Icarus**: The tale of Icarus, who flew too close to the sun with wings made of wax and feathers, serves as a cautionary symbol of hubris and the consequences of overambition. This myth has been referenced in various artistic and literary works, highlighting the peril of defying natural limits .
3. **The Green Man**: A symbol of rebirth and the cycle of growth, the Green Man appears in various cultures as a face surrounded by or made of leaves. In literature, he represents the connection between humanity and nature, often embodying themes of fertility and the natural world .
4. **The Dragon in Beowulf**: In the epic poem *Beowulf*, the dragon represents the inevitability of death and the

hero's final challenge. Its hoard symbolizes the transient nature of wealth and the futility of materialism .

5. **The Sirens**: In Homer's *Odyssey*, the Sirens are mythical creatures whose enchanting songs lure sailors to their doom. They symbolize irresistible temptation and the dangers of indulgence, serving as a metaphor for the perils of unchecked desire .

Common Themes in Mythological Imagery

- **Creation and Destruction**: Myths often explore the origins of the world and the forces that shape or destroy it.
- **Transformation**: Characters undergo physical or spiritual changes, symbolizing growth, decay, or renewal.
- **Fate vs. Free Will**: The tension between destiny and personal choice is a recurring motif.
- **Duality**: Many myths present opposing forces, such as light and dark, good and evil, chaos and order.
- **Sacrifice**: Heroes often undergo suffering or make sacrifices for the greater good, embodying themes of altruism and redemption.

Visual Representations in Art

Mythological imagery isn't confined to literature; it also plays a significant role in visual arts. Classical Greek pottery, Renaissance paintings, and modern sculptures often depict scenes from mythology, using symbolism to convey complex narratives and themes.

If you're interested in exploring how mythological imagery manifests in specific cultures, periods, or artistic forms, feel free to ask!

W

You said:

can you give me samples of mythic style imagery?

ChatGPT said:

Absolutely! Here are some evocative examples of **mythic-style imagery**—rich, archetypal, and symbolic—drawn from various mythologies and literary traditions. These images are designed to evoke timeless themes of transformation, heroism, fate, and the divine.

Dragon of Beowulf

A fire-breathing sentinel guarding treasure beneath the earth, its molten breath and ancient hoard symbolize the destructive power of greed and the inevitability of death. In *Beowulf*, the dragon's final battle with the hero marks the end of an era, embodying the clash between mortality and legacy.

Prometheus Unbound

A titan defying the gods to bestow fire upon humanity, Prometheus' eternal punishment—chained to a rock with an eagle eternally feasting on his liver—symbolizes the cost of enlightenment and the rebellion against divine authority. His myth has inspired works like Mary Shelley's *Frankenstein*, where the protagonist's overreaching ambition mirrors Prometheus's defiance.

The Green Man

A face formed from intertwining leaves and vines, the Green Man appears in various cultures as a symbol of rebirth, nature's cycles, and the interconnectedness of life. In literature, he represents the primal force of nature and the eternal return of spring.

Icarus' Fall

Wings crafted from wax and feathers, soaring toward the sun, only to melt and plummet into the sea. Icarus' tragic end serves as a cautionary tale about hubris and the perils of overreaching ambition. His myth has been referenced in various artistic and literary works, highlighting the peril of defying natural limits.

Orpheus and Eurydice

A lyre-playing poet descending into the underworld to reclaim his lost love, Orpheus' journey represents the power of art and the fragility of human desire. The moment he looks back, losing Eurydice forever, underscores the tension between hope and despair.

The Labyrinth

A maze of twisting corridors and dark passages, the labyrinth symbolizes the complexity of the human psyche and the journey toward self-discovery. Theseus' navigation of the labyrinth to defeat the Minotaur reflects the struggle to overcome inner demons.

Medusa's Gaze

A woman transformed into a monster with snakes for hair, whose gaze turns onlookers to stone, Medusa embodies the destructive power of beauty and the consequences of divine punishment. Her image has been used to explore themes of victimization and transformation.

Yggdrasil, the World Tree

An immense ash tree connecting nine realms, Yggdrasil represents the interconnectedness of all life and the cyclical nature of existence. Its roots delve into the underworld, while its branches reach into the heavens, symbolizing the unity of opposites.

Mjölnir, Thor's Hammer

A mighty weapon capable of summoning lightning and thunder, Mjölnir symbolizes strength, protection, and divine authority. Thor's wielding of the hammer in battles against giants reflects the triumph of order over chaos.

The Phoenix

A bird that bursts into flames and is reborn from its ashes, the phoenix symbolizes immortality, renewal, and the cyclical nature of life. Its myth has inspired countless works, representing the resilience of the human spirit.

If you'd like, I can help craft a piece of writing—be it a poem, short story, or scene—infused with mythic imagery. Just let me know the tone or theme you're aiming for!

Sources

You said:

can you do child poetry imagery

ChatGPT said:

Absolutely! Let's explore **childhood poetry imagery**—vivid, sensory-rich, and playful—designed to spark imagination and wonder. Children's poetry often uses:

- **Simple, sensory-rich language**: Engaging sight, sound, touch, taste, and smell.
- **Personification**: Giving life to inanimate objects or animals.
- **Repetition and rhyme**: Creating rhythm and musicality.
- **Whimsical themes**: Celebrating everyday wonders and adventures.

Animal Adventures

- *"The moon whispers secrets to the sleepy sea, while stars play hide-and-seek behind the clouds."*
- *"A squirrel in a sweater sips tea with a mouse, as the wind hums lullabies through the trees."*

Nature's Wonders

- *"Raindrops tap-dance on the windowpane, while flowers giggle in the garden breeze."*
- *"The sun wears a golden crown, painting the sky with hues of pink and lavender."*

Imaginative Play

- *"A dragon in pajamas reads bedtime stories to a teddy bear, as the clock winks at the moon."*
- *"Clouds turn into castles, and rainbows are slides for the giggling sun."*

Everyday Magic

- *"The clock winks and whispers, 'It's time for dreams,' as the night tucks in the sleepy street."*
- *"The kitchen hums a tune, as cookies dance on the counter, waiting for a nibble."*

These examples showcase how child poetry imagery delights in the ordinary, transforming it into the extraordinary through playful language and vivid imagery. If you'd like, I can help craft a personalized poem for a specific occasion or theme!

Sources

many examples and detailed explanation of the Iowa school of poetry and imagery

ChatGPT said:

Absolutely! The **Iowa School of Poetry** is a well-known tradition in American poetry, closely associated with the **University of Iowa's Writers' Workshop**, which is one of the most prestigious creative writing programs in the United States.

What is the Iowa School of Poetry?

- The **Iowa School** is not an official "school" with a strict manifesto, but rather a loosely defined group of poets who share certain stylistic and thematic traits.
- It emerged mainly in the mid-20th century, flourishing in the post-WWII era.
- Poets linked to the Iowa School often emphasize **precision, clarity, and vivid imagery**.
- There is a strong focus on **craft, form, and the poem as a careful construction**.
- The movement tends to avoid excessive abstraction or overt didacticism, favoring instead **concrete, sensory details**.
- Many poets connected with the Iowa School worked or taught at the University of Iowa's Writers' Workshop.

Characteristics of the Iowa School Poetry

1. **Imagery**:
    - The use of clear, strong, and often naturalistic imagery is central.
    - Poets focus on making images come alive with detail and vividness.
2. **Clarity and Precision**:
    - Language is straightforward but carefully chosen.
    - The poems are usually accessible and avoid unnecessary complexity.
3. **Craft and Form**:

   - Attention to meter, line breaks, and form is common.
   - Poems often have a quiet musicality and controlled tone.
4. **Emotional Restraint**:
   - Emotions are present but not over-explained or sentimental.
   - Feeling is often conveyed indirectly through imagery.
5. **Focus on the Ordinary**:
   - Everyday life, common objects, and natural scenes are frequent subjects.
   - The ordinary is shown as worthy of attention and poetic treatment.

Notable Poets Associated with the Iowa School

- **Mark Strand**
- **Donald Justice**
- **Charles Wright**
- **James Tate**
- **Jorie Graham** (though she evolved in other directions)
- **Philip Levine** (often linked to Midwestern themes and craft)

Examples & Explanation of Iowa School Poetry and Imagery

Example 1: Mark Strand — *"Keeping Things Whole"*

In a field
I am the absence
of field.
This is always the case,
but especially now
that, with my back to the world,
I am thinking of something else.

Explanation:

- The imagery is sparse but vivid. Strand uses the "field" and "absence of field" to evoke a strong visual and conceptual image.
- The poem embodies **emotional restraint**—the speaker is not overtly sentimental but evokes a subtle, meditative mood.
- The language is clear and precise; the metaphor is concrete and yet leaves room for interpretation.

Example 2: Donald Justice — *"Departure"*

When I left
I left nothing behind me
but the slow, steady shadow
of the slow steady sun.

Explanation:

- Justice's poem uses natural imagery ("slow, steady shadow," "sun") to evoke a mood of quiet departure and finality.
- The imagery is direct and tactile, inviting the reader to visualize the exact moment and feeling.
- The language is simple but carefully crafted to create a feeling of stillness.

Example 3: Charles Wright — *"Black Zinnias"*

The black zinnias on the window sill
have gone dark in the heat,
the wind stirring dust among the faded petals.

Explanation:

- Wright's imagery is detailed and sensuous—"black zinnias," "stirring dust," "faded petals" create a vivid and tactile scene.
- The poem uses the ordinary (flowers on a window sill) to explore themes of decay and time passing.
- The attention to sensory detail exemplifies the Iowa School's focus on the concrete.

Why Is Imagery So Important to the Iowa School?

- Imagery is the foundation of these poems; it grounds abstract emotions in tangible things.
- The poems often explore philosophical or emotional themes through concrete images rather than direct statements.
- The use of imagery fosters a connection between the reader and the poem's world, making the experience immersive and immediate.

Summary

The **Iowa School of Poetry** is known for:

- Clear, precise language
- Strong, vivid imagery drawn from everyday life and nature
- Emotional subtlety and restraint
- Careful attention to poetic form and craft

If you'd like, I can provide more poems with detailed line-by-line analysis or explain how the Iowa School influenced modern poetry in more depth! Would you prefer that?

Made in the USA
Monee, IL
12 November 2026

0d9386cb-5b62-4d3c-b8ea-7ecaf8999694R01